FLIGHT SIMULATION

Flight Simulation
Virtual Environments in Aviation

ALFRED T. LEE
Beta Research, Inc., Los Gatos, USA

ASHGATE

Published by
Ashgate Publishing Limited
Gower House
Croft Road
Aldershot
Hampshire GU11 3HR
England

Ashgate Publishing Company
Suite 420
101 Cherry Street
Burlington, VT 05401-4405
USA

Ashgate website: http://www.ashgate.com

British Library Cataloguing in Publication Data
Lee, Alfred T.
    Flight simulation : virtual environments in aviation
    1.Flight simulators 2.Air pilots - Training of
    I.Title
    629.1'3252'0113

Library of Congress Control Number: 2005923367

ISBN 0 7546 4287 9
ISBN 978 0 7546 4287 9

Reprinted 2006

Printed and bound in Great Britain by MPG Books Ltd, Bodmin, Cornwall

Contents

List of Figures and Table

Preface

In less than three decades following the first flight of an aircraft, engineers were applying the latest technology to simulate an aircraft on the ground. Since that time, the technologies necessary for realistic ground-based simulation of flight have grown dramatically in capability and complexity. Along with these technological developments has grown the realization that flight simulation design requires more than hardware and software engineering, it also requires an understanding of the piloting tasks and the complex human responses to the experience of flight. As a result, increasingly greater attention is now focused on a human-centered approach to the design and evaluation of piloted flight simulators.

This book has several objectives. The first is to describe the key component technologies of flight simulators and how they support the pilot's experience of flight and the pilot's performance of specific tasks. An additional objective is to provide some understanding of the capacity and limitations of the pilot in areas that are directly relevant to the design of flight simulation devices. How flight simulation technology is applied in the world of aviation is also described, as are studies that have examined the effectiveness of this technology in pilot training and evaluation. Finally, the limits of simulation technology are identified, as are technical advances that may significantly increase the utility and reduce the cost of flight simulators in the future.

The organization of chapters in the book is intended to meet the objectives stated above. The first five chapters of the book address key areas of simulation that are needed to create the experience of flight for the pilot. These include the simulation of the visual scene outside the cockpit, sound and communications simulation, simulating the sensation of aircraft motions, re-creating the handling qualities and control feel of an aircraft, and simulating the flight task environment. The latter chapter was added in the belief that the creation of a realistic task environment is essential if the higher level, cognitive skills of a pilot, such as workload management and decision-making, are to be trained and evaluated in a ground-based simulator.

Two additional chapters are devoted to the relationship between simulator fidelity and training effectiveness and to a discussion on how simulator fidelity should be defined and measured. Particular emphasis is placed on the importance of understanding how simulator design affects the experience of the pilot rather than simply how the technology can re-create the physical and functional characteristics of an aircraft cockpit.

The last three chapters address some inherent limitations of flight simulator technology that need to be understood by both those who design and those who use the technology. The chapter on advances in simulation technologies includes the potential use of intelligent computer-aided instruction in simulators. Finally, the substantial, and often overlooked, contribution of flight simulators in the field of aviation and human factors research is described in the last chapter.

It is hoped that these chapters will be successful in meeting the objectives of the book. It is also hoped that the book will help in creating new designs in flight simulation technologies and in stimulating ways in which the limits of the technology might be overcome. Flight simulation plays a vital role in both the training and evaluation of pilots. Modern civil and military aviation depend heavily on simulation technology not only for training, but also in accident investigations, in advanced aircraft research, and in understanding the interaction between pilots and the flight environment. The role of flight simulation in aviation is likely to increase in importance as the complexities of modern aviation systems make them increasingly indispensable for both pilot training and research. Improving the design of flight simulators and understanding how they affect the pilot's experience of aircraft flight are vital if that role is to be fulfilled.

Alfred T. Lee, Ph.D.

List of Abbreviations

3-D	Three Dimensional
3-DOF	Three Degrees of Freedom
6-DOF	Six Degrees of Freedom
ACARS	ARINC Communications and Address Reporting System
AGARS	Advanced General Aviation Research Simulator
AOI	Area of Interest
ARTCC	Air Route Traffic Control Center
AFS	Aerosoft Flight Simulator
ASOS	Automated Surface Observation Systems
ATC	Air Traffic Control
ATD	Aviation Training Device
ATIS	Automated Terminal Information System
AV-MASE	Air-Vehicle Modeling and Simulation Environment
AWOS	Automated Weather Observation System
BC	Back Course
CAMI	Civil Aviation Medical Institute
CGI	Computer Generated Imagery
C_{obj}	Object Contrast
COTS	Commercial Off-The-Shelf
CPT	Cockpit Procedures Training
CRM	Crew Resource Management
CRT	Cathode Ray Tube
CTA	Cognitive Task Analysis
CTAF	Common Traffic Area Frequency
FAA	Federal Aviation Administration
FFS	Full Flight Simulator
FMS	Flight Management System
FOV	Field of View
FOV_H	Horizontal Field of View
FOV_V	Vertical Field of View
FTD	Flight Training Device
GAT	General Aviation Trainer
GPS	Global Positioning System
GPWS	Ground Proximity Warning System
IATA	International Air Transport Association
ICAO	International Civil Aviation Organization
ILS	Instrument Landing System
IOS	Instructor Operator Station

LΔ	Difference between Luminance of Object and Luminance of Object Background
L$_{bg}$	Luminance of Object Background
L$_{max}$	Maximum Luminance
L$_{min}$	Minimum Luminance
LCD	Liquid Crystal Display
LOC	Localizer
LOFT	Line-Oriented Flight Training
LOS	Line-Oriented Simulation
LOS$_{obs}$	Observer Line of Sight
MLS	Microwave Landing System
NASA	National Aeronautics and Space Administration
NASA TLX	National Aeronautics and Space Administration Task Load Index
NDB	Non-Directional Beacon
PC	Personal Computer
PCATD	Personal Computer Aviation Training Device
SAS	Simulator Adaptation Syndrome
SRS	SIMONA Research Simulator
TCAS	Traffic Collision and Avoidance System
TER	Transfer Effectiveness Ratio
TOT	Time On Task
TRACON	Terminal Radar Control
TTS	Text-To-Speech
TWEB	Terminal Weather Enroute Broadcast
VHF	Very High Frequency
VMS	Vertical Motion Simulator
VOR	Very High Frequency Omnidirectional Radio

Chapter 1

Visual Scene Simulation

Introduction

The simulation of the visual scene that presents itself through the window of an aircraft has been one of the most significant technical challenges in the development of flight simulation technology. It is, in fact, one of the elements of flight simulators that did not begin to see to major progress until the 1980s and the onset of rapid growth in computer processing power. Prior to that decade, highly detailed visual scene simulation was largely delivered by the scanning of cameras over detailed terrain models.[1] Early attempts at scene simulation by computer-generation resulted in sparsely detailed images, which were of very low resolution. In the decades since the explosive growth of computer technology began, the ability to display real-time imagery over a large field-of-view (FOV) has become commonplace in modern flight simulators. This chapter explores the technology of visual scene simulation and its role in flight simulation as well problems in the design and implementation of the technology.

Aircraft Control by Visual Reference

Flying an aircraft[2] by use of the visual scene provided out of the cockpit window can be divided into three basic parts: flight control, navigation, and collision avoidance. Positive flight control can be further divided into attitude, speed, and altitude control. Control of aircraft attitude for the pilot is in three basic dimensions: pitch, roll and yaw. A simplified example of these three control axes in relation to the visual scene is seen in Figure 1.1. An aircraft pitches up and down when rotated about its y or lateral axis. Pitch attitude cues are derived from movement of the visual scene elements vertically up and down the aircraft wind screen. The most important visual scene element to the pilot using the visual scene

---

[1] Early image generation systems were composed of detailed terrain models and gantry-mounted cameras that moved across them in accordance with the simulated aircraft's attitude, speed and altitude. A version of this system was used to train astronauts in the U.S. space program's early years.

[2] Aircraft flight references in this book will generally refer to basic fixed-wing, flight operations unless otherwise indicated. Special uses of aircraft, such as helicopter or military operations, may significantly affect some elements of basic simulator design requirements.

for attitude control is the visual horizon. Even the most sparsely detailed visual scene can provide the necessary cues to control of aircraft pitch if a visible horizon is available. Indeed, provided one knows which part of the scene is sky and which part is ground, a visual scene devoid of all details *but* the visual horizon is adequate for aircraft pitch control.

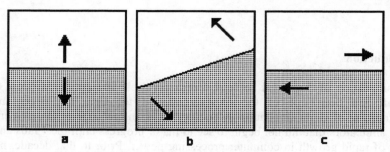

Figure 1.1 Visual Cues to Pitch (a), Roll (b), and Yaw (c)

Aircraft control in the roll axis is also dependent, to a large extent, on the existence of a visible horizon. Roll or rotation about the aircraft's longitudinal or *x* axis rotates the visual horizon around the point where the aircraft is headed. The angle of the visual horizon is the aircraft's angle of bank and the rate of rotation is the aircraft's rate of roll. Early simulator's with primitive visual simulation systems often provided only the visual horizon represented by a line segregating two halves of white (for the sky) and black or gray (for the earth). This was adequate for providing limited pitch and roll control provided the aircraft's pitch attitude was not so extreme as to completely void the display of the visible horizon. In these situations, if no artificial visual horizon was available, loss of control was likely (just as it is in the aircraft). The third axis of aircraft control which uses external visual reference, control of the yaw or *z*-axis, can be achieved only by reference to an object or reference point which can be displaced horizontally, left or right, across the windscreen. A simple line segregating earth and sky is therefore not adequate to provide feedback for control in the lateral axis. The image needs to include an object or other detail that will move in conjunction with aircraft yaw or rotation around the *z*-axis. Translational motions along the *z*-axis also result in horizontal displacements, though such motion is minimal in fixed wing aircraft (but may be much greater in other aircraft such as helicopters). Without these additional visual scene details, early visual simulations were extremely limited in the usefulness of the visual information they could provide and, therefore, in the use of the simulator for training aircraft flight control by means of visual reference.

Increased object detail in visual scene simulation is essential when an aircraft operates at altitudes at less than a thousand feet above ground level. At these heights, ground reference cues become increasingly important in the judgment of height and in the judgment of both rate of altitude change and the angle at which the aircraft is approaching the ground. Two essential elements of visual scene simulation play a role under these conditions. The first of these elements is the display of individual objects in the visual scene and the second is the display of details within each of the displayed objects.

Object details such as size are essential in order to provide the pilot with valid and reliable cues to depth and distance and to the *changes* in depth and distance that support aircraft control, particularly control of airspeed and control of altitude when close to the ground. The importance of accurate object rendering is related to the basic human visual processing of object details. Objects in our world typically have a known size. These include natural terrain elements such as trees as well as cultural features such as highways and houses. Because the objects change in perceived (or retinal) size as the distance to them changes, the pilot interprets the difference in perceived size as a change in depth or distance. This is possible because an object's size is normally invariant and changes in its size can only be due to changes in the distance of the object from the pilot. This *size-distance* relationship of objects is a powerful visual cue for the interpretation of height above ground, for the distance from another aircraft, for the rate of change in altitude and for many other piloting tasks that occur close to the ground or to other airborne objects.

Other visual processes of the pilot are also supported by accurate rendering of object details. These include object *detection* as well as object *identification and classification*. Object detection is determined by the retinal or perceived size of the object, the relative contrast of the object against its background, and the position of the object in the pilot's visual field. Identification of an object occurs when sufficient object detail is available in order for the object to be classified in some way. Often this is a matter of the pilot's ability to perceive key classification characteristics of the object that allow identification to take place. For example, classifying an airborne object as an aircraft involves the ability to determine whether the object's shape and size corresponds to any known aircraft type. Further classification processes may include determination of the aircraft's *aspect ratio*, that is, the direction of another aircraft's flight path relative to one's own. These visual detection, identification and classification processes are critical to flight tasks such as airborne collision avoidance and, in military aircraft operations, to air combat maneuvering. Object detection and discrimination processes are also used for objects on the ground in those flight applications where ground-based target acquisition and classification are important. Discrimination of ground targets is an important tool in pilotage where navigation is dependent upon the identification of key ground objects such as roads, lakes, rivers, buildings, and other features which can be used in conjunction with maps and charts to identify key waypoints and visual approach points to airports and in locating controlled airspace.

It should be evident from the above descriptions of piloting tasks that accurate rendering of object details in visual scenes may have a significant impact on the utility of flight simulators. Distorted rendering of objects in flight simulator visual displays or the absence of key image details can not only limit the usefulness of flight simulators but, perhaps more significantly, alter the way in which pilot's learn key flight skills.

Monocular Cues to Depth and Distance

All objects in a simulator visual scene are displayed on a two-dimensional surface. They therefore lack the normal binocular distance cues that are provided in real life, 3-dimensional (3-D) imagery. These distance cues are the product of binocular disparity effects resulting from the display of slightly different images to each of the two eyes. However, these binocular disparity cues are only significant at ranges below 2 m. With rare exceptions, aircraft do not operate this close to other objects either in the air or on the ground. For this reason, visual scene simulations in flight simulators will normally provide only *monocular* cues to depth and distance.

Linear Perspective

One of the most compelling monocular cues to distance is linear perspective. Simply put, linear perspective is produced by the visual merging or convergence of parallel lines as the points on the lines recede from the viewer. The more obvious examples of linear perspective cues for the pilot are provided by airport runways and taxiways. When a third dimension, altitude, is added, the linear perspective of the resulting object forms a powerful visual cue, particularly during the approach and landing phase. This cue varies in relation to the changes in an aircraft's approach angle and distance from the runway (see Figure 1.2).

Improper rendering of the runway object in a simulator visual scene can occur for a variety of reasons, including poor display resolution or inadequate vertical display field-of-view. In poor resolution displays, display aliasing can introduce artifacts into the visual scene such as foreshortening of the departure end of the runway so that the runway edges appear to join in the distance. Lack of vertical field-of-view can also cut off the runway end of long runways and thus fail to provide a rendition of the runway object shape as it would appear in the real visual scene.

Aerial Perspective

Typically, flight simulator image displays provide visual scenes that encompass great distances both horizontally and vertically. In real world operations, phenomena such as water particles, smoke and other atmospheric contaminants can dramatically alter what the pilot actually sees at these distances. Collectively, these phenomena create a visual effect called *aerial perspective*. As with linear perspective, aerial perspective provides a strong cue to distance and depth. This is because atmospheric particulates scatter and diffuse light emitted or reflected from an object. The scattering and diffusion of light increases as the distance of the observer from the object increases. Furthermore, the color (hue) of objects becomes increasingly desaturated or 'washed out' with increasing distances between the pilot and the object.

Figure 1.2 CGI Image of Runway on Final Approach
(Image Courtesy of Adacel, Inc.)

While aerial perspective cues will tend to correlate with the distance at which objects can be seen, the effect can be dramatically altered by the position and intensity of light sources. An example of aerial perspective in a CGI display system is shown in Figure 1.3. Note how the aerial perspective increases with distance in the image and eventually obscures the visible horizon.

Texturing and Texture Gradients

Other monocular cues to depth and distance are available to the pilot in real life. Among the most powerful of these is image texture or, more accurately, the texture gradient. A texture gradient is simply the change in density of texture elements as a function of distance from the viewer. As the density of texture increases as a direct function of distance from the viewer it provides the pilot with a cue to depth and distance. Objects in the distance which have a texture gradient in the foreground of the visual field will, therefore, always be perceived as further away than an object of the same size with less or no texture gradient in the foreground. Texture gradients are also useful cues to judging the relative slant or slope of terrain. Typically, flat terrain surfaces in, for example, forested areas will reveal the typical texture gradient characteristic as the distance from the observer. As the slope of terrain changes so will the texture gradient of the surface objects on the terrain.

Figure 1.3 Aerial Perspective in a CGI Display
(Image Courtesy of Adacel, Inc.)

Texture gradients may have their greatest utility in aiding a pilot during the landing flare. It is at this stage of the landing process where judgment regarding the aircraft's height above ground and the aircraft's rate of sink to the ground is most important. The landing flare is one of the most difficult tasks for student pilots to master. It is also a skill that can deteriorate rapidly if not refreshed either in a flight simulator or in an aircraft. Providing texturing of runway surfaces as it occurs in the real world visual scene is therefore essential to the successful training of the landing flare maneuver.

The absence of texture in a flight simulator display of an airport runway can lead to the development of *compensatory skills*. These skills are developed in flight simulators as a means of overcoming the limitations or improper design of a simulator. While many such skills may be harmless, others may inadvertently introduce safety problems into flight operations. In a study by Mulder, Pleisant, van der Vaart, and van Wieringen (2000), trainee pilot performance in the landing flare was compared with and without runway texturing in a flight simulator. Pilots in the non-texture condition developed the strategy of using runway edge markings as a means to judge height above the ground while pilot trainees in the texture condition used the runway's texture as a flare cue instead. For pilots in the non-texture condition, developing the strategy of controlling the landing flare by using the distance between runway edges would only be useful if all runways are of the same width. As runways vary widely in width in the real world, pilots in the non-textured display condition acquired a skill that will often be inappropriate in real world operations. When applied to real aircraft landings, these pilots may initiate the landing flare too early or too late. In this example, the absence of runway texture in the simulated visual scene required the pilot to develop a compensatory skill which could adversely affect aircraft safety.

Other Monocular Cues to Depth and Distance

A number of other monocular cues to depth and distance are available to the pilot. These include object overlap or occlusion, where objects nearer to the pilot overlap portions of objects further away. The occluded object will always be interpreted by the pilot as being further away than the occluding object. Object occlusion occurs more often in ground operations since there are likely to be more objects in the visual field that can provide the effect. Large numbers of ground objects in the visual field will provide greater opportunities for occlusion or overlap to occur and will thus provide a greater sense of depth to the visual scene simulated. Airborne operations also benefits from this cue in a visual scene. Successive overlapping of multiple objects often occurs when flying over high-density urban scenes, forests, and mountain ranges.

Motion in Depth

Motion in depth can also serve as important distance and depth cues. One monocular motion cue to depth is *motion parallax*. When objects in the visual scene are at different distances from the observer's line-of-sight (LOS_{obs}), the movement of the objects will differ relative to one another as the observer moves through the scene. This effect is produced because of the relative differences in angular motion in objects at different distances from a moving observer. Figure 1.4 provides an illustration of this effect. In this figure, the pilot is viewing the scene out the right side of the aircraft while the aircraft is moving forward. The speed of objects in the scene varies as a function of their position relative to the LOS_{obs} as illustrated by the increasing length of the arrows. Objects above the LOS_{obs} move in the same direction of the aircraft motion and opposite the direction of those objects below the line of sight.

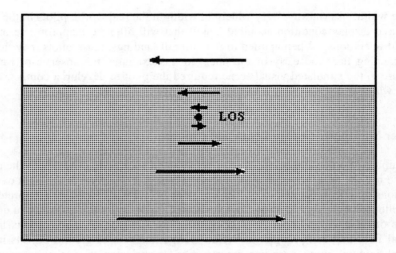

Figure 1.4 Motion Parallax in a Visual Scene

Motion parallax cues may play an important role in providing distance cues to pilots in low level flight and ground operations. As the effect is produced by the relative motion of objects placed at different distances in the visual field, the larger the number of objects in the visual scene, the more likely they will aid in producing the motion parallax depth or distance cue. This is particularly true if other monocular depth or distance cues, such as texture, are not available in the visual scene.

Motion flow patterns can interact with other cues in the visual scene to create both a sense of depth and distance in the scene. Optical flow patterns created by the motion of objects in the scene can occur in a variety of forms. In Figure 1.5, an optical flow pattern is generated when the LOS_{obs} and aircraft heading is at the center of the visual horizon. In this case, objects are streaming past in an optical flow pattern radiating from the LOS_{obs} at varying angles. Both the particular patterns and speed of optical flow provide important cues to aircraft visual flight control. Such flow patterns are dependent on the density and position of objects in the visual scene. Optical flow patterns are particularly important for visual scenes intended to simulate aircraft ground operations, takeoff, visual approach and landing and low level flight maneuvers.

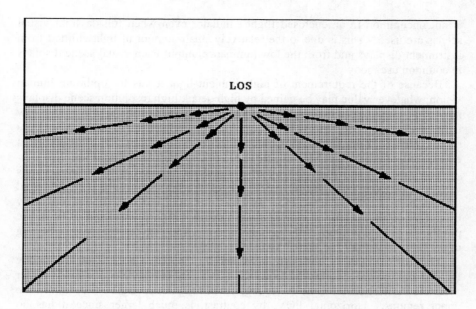

Figure 1.5 Optical Flow Patterns in Aircraft Forward View

Visual Scene Simulation Technology

Limitations in the Human Visual System

Before discussing the technology of visual scene simulation in detail, it is necessary to note some of the limiting characteristics of the pilot's visual system. With advances in simulation technology increasingly so rapidly, simulator designers need to consider the limits of the human visual system. Providing complex and costly visual scene simulation technology that exceeds the capability of the human visual system is something that should obviously be avoided. For this reason, a brief description of some limitations in the human visual system is presented here.

The loss of pilot visual acuity in low light levels is particularly important for simulator display design as the ambient light in most enclosed simulators is relatively low when compared with that available in the aircraft cockpit during daytime operations. The ambient light available to the eye from all sources in the simulator (including the visual display) is typically much less than that available in the aircraft under daylight conditions and even less than might be expected in normal office settings. In fact, a total ambient illumination of only 50 to 100 Lux[3]

---

[3] By comparison, aircraft cockpit ambient illumination during daylight operations may be 1000 Lux or higher

is not uncommon in an enclosed flight simulator even when full daylight visual scenes are used. This is due to the relatively small amount of light emitted from instrument displays and from the low luminance output from visual scene displays in common use today.

Because of the requirement of high ambient light levels for optimum human vision, whatever value may be gained by high detail, high resolution scene displays may be lost because of the poor operating characteristics of the pilot's eyes at low light levels. In order to avoid this problem, simulated daylight visual scenes should provide a highlight luminance of at least 300 cd/m^2. This assures that enough ambient light will be available to provide for normal daylight vision.

The total field of view (FOV) of a simulator image display system varies widely depending upon the intended use of the simulator. The human visual system, however, has a fixed FOV which is produced horizontally and vertically by the two eyes working in combination. The resulting instantaneous FOV of the pilot's visual field is about 180 deg horizontal and 130 deg vertical. About 140 deg of the horizontal FOV are shared by the two eyes.

The vertical FOV of both eyes is limited by the ocular bone structures above the eye and the cheekbone structures below the eye. This results in a much larger available vertical FOV in the lower regions of the pilot's visual FOV than in the upper regions. Horizontal FOV, by contrast, is much larger since it has no structures that might restrict it in the temporal regions. The nasal periphery, however, of each eye is partially blocked by the nasal bone structures. The extensive horizontal FOV in the temporal region exposes the pilot's visual field to optical flow fields and individual object motion in the far visual periphery.

The eye does not have the same resolving power across its entire FOV. For daylight vision, about 80 percent of the resolving power of the eye lies within ten deg of its center and tapers off rapidly into the visual periphery. The maximum resolution of 1 arc min only exists in the central region of the eye within 1 to 2 deg of the center of the eye. At between 2 and 10 deg of eccentricity, visual acuity drops to about 5 arc min and from 10 to 20 deg it drops to around 10 arc min. The pilot's visual FOV beyond 10 deg from the center of the eye or *fovea* is therefore poorly equipped to detect and recognize objects, although this area is sensitive to motion in the visual scene. Since the nominal 1 arc min resolution occurs only within about 1 to 2 deg of the center of the eye, much of what is displayed at this level of resolution on a simulator visual display will be of value only when the pilot is looking directly at it.

Many of the limitations in FOV and resolving power of the human visual system are overcome by eye and head movements by the pilot. This allows a dramatically expanded FOV while allowing the high acuity central region of the eye to be placed on object details of interest. However, in some tasks, such as landing and takeoff, the pilot's head and eye movements will be much more limited as these tasks require more constant eye and head positions.

In the description of visual scene simulation technology that follows, the limitations of the pilot's own visual system should be kept in mind. It is likely that, as technology improves the capability to generate and display visual scenes, the pilot's visual system, and not technology, will limit what needs to be displayed.

Image Generation

In modern digital simulation, Computer Generated Image (CGI) scene architecture begins with several basic components. The first of these is the visual gaming area, or simply 'gaming area' which is composed of a pre-defined segment of terrain of a known size, for example 25 nm x 25 nm or 100 nm x 100 nm. The gaming area size is determined by the amount of detail capable of being displayed by the imaging generating system at a given time. This is largely a function of the processing and memory power available to the simulator computer system. For a terrain of a given complexity, larger and more powerful systems are able to handle larger gaming areas. The advantage of large gaming areas is that they can provide smooth, seamless imagery across the simulated terrain or airspace. With small gaming areas, imaging systems are burdened with the repeated loading and unloading of images into memory, which may result in erratic imagery display or the temporary loss and reappearance of scene components. As computation power has increased, the size and complexity of gaming areas has increased accordingly. Many of the problems related to gaming area image display which plagued earlier systems have, for the most part, been eliminated. The design of gaming areas in modern systems is now more a function of how much actually needs to be displayed for a given simulation scenario and less about computational limits. Nonetheless, computational as well as cost and simulator utility tradeoffs will affect the realism and utility of the gaming area generated. To answer these questions, the visual scene simulation designer needs to know how visual information is used in the flight task as well as the limitations of human visual information processing.

Rendering Objects in CGI

In modern CGI systems, 3-D rendering of objects is accomplished by computing and drawing polygons, the vertices of which are individually addressable picture elements or pixels. Typically, scene complexity is defined in terms of the number of polygons that can be displayed at any one time. The larger the number of polygons, the greater the computational power needed to render the visual scene accurately. For this reason, it is often the case that image generation systems are rated by the number of polygons that can be drawn in a given visual scene. Defining the objects needed and the key characteristics of these objects that need to be seen within the gaming area will significantly affect the number of polygons that the simulation must draw in each new frame of the scene.

The *update rate* of a CGI system is driven largely by the minimum number of times the display image or frame needs to be updated in order to provide the pilot with the illusion of motion of the objects displayed. It is possible to calculate the minimum update rate required by using the following equation derived from Padmos and Milders (1992):

$$U = A/15$$

where:

U = minimum update frequency in frames per sec
A = object angular speed in arc min per sec

Object angular rate is divided by 15, the displacement (in arc min) needed per frame that will provide the illusion of continuous motion. Update rates are typically set for a given display system at a rate necessary to provide the illusion of smooth object motion. A typical update rate for flight simulators is 30 frames per sec, but the actual rate is dependent on the visual scene simulation requirements for object motion. In some applications, much higher update rates may be required. For example, evidence exists that suggest that high speed, low level flight may require update rates as high as 60 frames per sec (Zindholm, Askins, and Sisson, 1996). The high update rates necessary to provide the illusion of smooth object motion can tax the computational power of even the most powerful image generation system. As an illustration, a single display system of 1000 x 1000 pixels required for training low level, high speed flying would need a system capable of updating 1 million picture elements sixty times each second. This represents a total of 60 million pixel updates for each second of visual simulation.

Fortunately, some savings in image generation capability is afforded by the fact that complete object details do not need to be generated at all times. The human eye cannot resolve very small objects or object details at long ranges and some scene details are always hidden from the viewer in any case. For these reasons and others, image generation systems do not need to draw every element of every object continuously. Imaging systems will typically generate increasingly complex object details only at closer object distances while rendering objects more crudely at greater distances where certain object details cannot be seen. If done properly, the object details are seen only when they are needed to be seen and the transitions where object details appear and disappear will be smooth and not distracting. The decision as to which object details should be generated at a given distance, however, should not be made arbitrarily. The decision should be based on whether such details would be visible by the human eye at the simulated object distance, whether the image display system will have the resolution needed to display the object details, and whether the object details serve a specific training or research need.

Object Animation

Animation of CGI objects has also become possible with the advent of advanced computers. Individual objects such as aircraft and ground vehicles can not only populate the scene, they can also move independently and automatically in accordance with pre-programmed scripts. These autonomous and semi-autonomous objects can move in visual scenes in pre-determined fashion triggered by scenario events or by a human operator. The object's programming allows automatic coordination with other autonomous objects and other elements of the scenario. For example, busy airports typically have a variety of airborne and ground-based aircraft traffic in various stages of taxi, takeoff, approach and landing. All of these objects can be animated in a way that simulates realistic behavior of such operations and does so without the direct control of the simulator user. Aircraft objects can also be coordinated with simulated ATC communications, so that there is a correct correspondence between simulator ATC instruction and the behavior of the autonomous aircraft object in the simulated visual scene.

Generating Texture in Displays

In early simulator visual display development, the display of texture proved to be nearly impossible owing to the extreme limitations placed on the amount of detail that could be provided in a simulated scene. The absence of texturing gave the visual scenes a 'cartoonish' appearance. More importantly, pilots had to find other visual cues to judge height or develop skills to compensate for the lack of these cues. In real life, texture gradients may be composed of individual elements numbering in the thousands or millions. Generating large textured surfaces such as runways was simply out of the question for the computers of early CGI systems.

The solution to this problem came by rethinking the problem of a texture gradient, not as a collection of individual objects as occurs in the real world, but as a more or less random pattern of varying intensities of light. Rather than generating individual elements of the textured surface as objects, a mathematical algorithm was used which described the light pattern on a given surface. The texture fill algorithms also allowed the surface fill of an object in a CGI display, such as a runway, to change as a function of the change in slope or slant of a displayed surface. Texture fill or texture tiling is found in virtually all modern simulator visual displays, which can now mimic the visual characteristics of a true textual gradient without the need to draw large numbers of individual objects. An example of texturing in a CGI display is shown in Figure 1.6.

Figure 1.6 Runway Texturing in a CGI Display
(Image Courtesy of Adacel, Inc.)

As a surface fill, texture tiling contains no individually rendered 3-D objects. Thus, if texturing is used to simulate tree cover of a hill, none of the component elements of the texturing will reveal the normal features of 3-D objects that often come to view at different viewing angles. This makes the use of individual texture elements as cues[4] for height or distance judgment less effective then they might otherwise be in the real world visual scene. Texture fills, while a valuable tool in improving visual scene realism, have limits which need to be understood when training or evaluating pilots for low level flight operations.

Related to the problem of texturing is the problem of filling large areas of the visual scene gaming area with image details approximating those of the real image. To reduce the cost of developing realistic scenery in simulation, the use of different categories of texturing have now been developed to automatically fill large areas of simulated visual scenes. For example, there are texturing bitmaps for forest, desert, and urban scenes as well as texturing bitmaps for runway surfaces, taxiways, and airport ramp areas. These bitmaps can be changed to reflect changes in seasons. To improve the realism of filling very large scenes even further, aerial photos and satellite images are now being employed in flight simulator visual

---

[4] The term 'cue' is used here to describe stimuli that have specific meaning for a pilot. The term 'stimuli' is used to describe sensory inputs that have no specific meaning for the pilot.

scenes. With photo-realistic techniques where the fill is based on actual photos, the resolution of the original photos determines what will be seen at a given altitude in the simulator. A satellite image of a resolution of 5m/pixel, for example, will look realistic in the simulator at several thousand feet of altitude or higher but will become blurred or 'pixilated' at lower altitudes. Use of photo images as surface fills at very low altitudes will likely require photo image resolutions in the sub-meter/pixel range to provide realistic visual scene simulation at these low altitudes.

Display Contrast and Luminance

The maximum luminance (L_{max}) and minimum luminance (L_{min}) of a visual display are calculated in order to determine the maximum object contrast available to the pilot in the visual display. Object contrast is particularly important for object detection and discrimination task performance by the pilot and therefore needs to be accurately simulated. Typically, L_{max} is the luminance value of the highlight brightness of the display where highlight brightness is measured from the white elements of the display test pattern. L_{min} would be the minimum luminance of the display and is generally measured from the black elements of the test pattern.

The contrast of an object from its background is calculated relative to its immediate background not the minimum luminance available in the display. The contrast for an object is therefore an incremental measure relative to its immediate background. The formula for object contrast is:

$$C_{obj} = \frac{L\Delta}{L_{bg}}$$

where:

$L\Delta =$ difference between the object's luminance and the luminance of the object's immediate background
$L_{bg} =$ luminance of the object's immediate background

An object brighter than its background has a positive contrast, or positive object polarity; an object with less brightness than its background has a negative object polarity. Higher object contrast (positive or negative) is needed to detect objects of smaller size, so simulator displays which have very high resolution generally need to have the capability of providing a greater range of object to background luminance contrast. Additionally, contrast sensitivity of the eye declines as a logarithmic function of ambient light available so the ambient light available to the eye from within the simulator must be high enough to assure adequate contrast sensitivity in order for small objects or object details to be seen. The unfortunate consequence of this is that simulators that have very low ambient light levels may effectively neutralize the benefits of high resolution image displays when simulating day visual scenes. As low ambient luminance levels also affect pilot color perception, low display luminance levels may also reduce the pilot's ability to correctly process colors in the visual display. To assure that both contrast sensitivity and color sensitivity of the pilot is not impaired in day visual scene simulation, high ambient light levels in the simulator are needed. As much of the light available in the simulator is from the visual display system, the visual system's luminance level needs to be high. As noted earlier, a luminance level (highlight brightness) of at least 300 cd/m^2 is recommended for simulator visual displays to assure that adequate light levels in the simulator are available at the display design eyepoint.

Display Resolution

An essential ingredient in the display of flight simulator visual scenes is image resolution. A scene generation system may be fully capable of providing all of the information needed, but without adequate display resolution the generated information is of no value. Computer-Generated Imagery (CGI) display resolution is generally measured in terms of *addressable resolution*. Addressable resolution is the number of pixels that can be independently addressed by the computer for a given visual channel. A more important measure of display resolution is *spatial resolution*. Spatial resolution is the number of pixels for a given linear dimension of a display. For example, a display which has 1024(h) x 768(v) pixels and a viewable horizontal dimension of 100 cm has a spatial resolution of 10.24 pixels per cm (ppcm). A single number is typically used to describe the spatial resolution for an image display system. Since it more accurately conveys the amount of detail a system can actually display, spatial resolution should be used instead of addressable resolution when specifying the image resolution of a display.

Since the image detail that the pilot can actually see is what really matters, a third display resolution specification is needed. This is the effective display resolution or simply *effective resolution*. The effective resolution of a display system is the resolution measured at the display design eyepoint. This eyepoint has a known distance from the display surface for any display and is the design point for which all of the display generation geometry is calculated. In order for a given object to be displayed at the correct size for the simulated viewing distance, the object size as displayed on the display surface must be combined with the distance of the display from the design eyepoint in order that the displayed object will be viewed at the same visual angle as it would be in the real world visual scene.

The effective resolution of the display is determined by the visual angle subtended by a single displayed pixel at the pilot's eyepoint. For example, if the single pixel of a display is 0.1 cm in diameter at the display surface and the pilot's eyepoint is 1 m from the display surface, the effective resolution is calculated as follows:

$$R_e = 57.3 \ H/D$$

where:

R_e = Effective resolution (visual angle subtended in degrees)

H = Pixel diameter or pixel height or width

D = Distance of the displayed pixel from the design eyepoint

For this example, the display's effective resolution is .057 deg or 3.42 arc min per pixel. The reason for the inclusion of a measure of effective resolution is due to the fact that there are many different methods of displaying scenes in flight simulators which may present the same spatial resolution at very different distances from the design eyepoint. Of all the measures of display resolution, the display's effective resolution is the most important for determining the level of detail a pilot will actually be able to see in the simulator's displayed image.

Determining Display Resolution Requirements

It might be a simple expedient to simply require all simulator display effective resolutions to be set at the nominal 1 arc min noted earlier. However, cost is a factor in setting the effective resolution of a display system, particularly if the system needs to cover a very large effective FOV. Getting the effective resolution to match the maximum visual resolving power of the human eye (1 arc min) simplifies the requirements specification, but may be cost-prohibitive. Fortunately, in many applications, this very high level of display resolution may not be needed. An alternative means to setting effective resolution equal to the maximum resolving power of the human eye is to determine the object or object details that need to be seen at a given range to support a particular training or research objective. The cross-section of the object or of an object detail and the range at which that object or objects detail needs to be seen are combined to provide the maximum effective resolution for that display system.

For example, if it is desired to train pilots to detect and avoid small aircraft at a great enough distance to train collision avoidance maneuvers, both detection and identification of the other aircraft's aspect angle is needed. The first task, detection, is most difficult when the aircraft cross-section is at its smallest, as occurs when the aircraft is flying directly at the pilot's own aircraft. For civil aviation, collision avoidance for small aircraft is likely to be the most important. Assuming the frontal cross-section of the fuselage for a small aircraft is only about 140 cm in diameter and it is desired to train a pilot to detect it at a range of two nautical miles, then an effective resolution of 1.43 arc min per pixel is needed for the display system. This resolution is more than that needed to identify the aircraft's aspect angle so it represents the maximum resolution needed for this particular display system.

Using this technique, the display system's maximum resolution can be specified at far less than the nominal 1 arc min per pixel while still providing the necessary effective resolution for the training or research purpose at hand. Since there is typically only one value for a system's maximum effective resolution (but see Ch. 9 for exceptions), a careful analysis of the end-use requirements of the simulator needs to be conducted to assure that the resolution is adequate.

Display Color

Color perception plays only a limited role in flight operations and therefore is less important than other display factors that have been discussed. Other than a very limited range of colors associated with natural scenes, much of pilot color perception is important only for those specific colors required in aviation operations. Part of the licensing of airmen includes their ability to discriminate a limited set of specific colors deemed of importance to flight safety. These are the so-called aviation colors, such as 'aviation red', 'aviation blue', and 'aviation green'. The precise characteristics of these colors are designated by regulatory authorities such as the Federal Aviation Authority (FAA) or the International Civil Aviation Organization (ICAO). They are used in both aircraft and airport lighting and are standardized throughout all aviation environments. Aviation red is used in navigation lights on aircraft, on visual approach lighting, and on runway lighting. Aviation green is also used in aircraft navigation lighting and for tower beacons while aviation blue is used for taxiway edge lighting. Amber lighting is used at the intersections of runways and taxiways. These and other designated aviation colors must be rendered accurately in image display systems since these colors transmit information important to the simulation of safety-critical, scene components.

What is typically described as 'color' in an image display or in the real world by the casual observer is actually a combination of hue or dominant wavelength, luminance or brightness, and saturation. The first of these, hue, is what defines color, which is the dominant wavelength of the light visible from the displayed object. For example, the color 'blue' may be assigned to an object color which has a dominant wavelength of around 650 nanometers. The dominant wavelength of emitted or reflected light from an object must be between 400 and 700 nanometers in order for it to be seen as having color by the human eye. Wavelengths below this range (infrared) or above it (ultraviolet) cannot be seen by the human eye. Accurate color perception only occurs during daylight viewing conditions.

In CGI systems, different colors are displayed by manipulating bits assigned to each pixel in the display. A system color palette is limited by the byte depth of the system. A system that has an 8-bit byte depth can display 256 colors, for example, while a 24-bit system can display more than 16 million colors. More important than the actual number of colors displayed is the accuracy with which such colors are rendered. Color rendering accuracy is a function, not just of the size of the system color palette, but also of the display system components, especially the image display system. Accurate color rendering is particularly important for the display of airport and aircraft lighting as the color of these lighting systems must conform to regulatory requirements for color accuracy in real life operations. For example, 'aviation red' needs to be rendered by the display system to conform to the particular wavelength established for this color by regulatory authorities. Simple reliance on color palettes provided by color monitor software systems are not adequate to meet the color accuracy requirement as the image display systems can alter substantially the color actually displayed to the pilot. Color spectrometers should be used to verify the actual color rendered by a visual display.

Image Display Systems

While image generation creates the objects, texturing and other elements of the image, it is the image display system that ultimately presents the pilot with the simulated visual scene. Three general types of image display systems are available for flight simulator use: projection, direct view, and direct view/collimated. Each of these systems has significant advantages and disadvantages in terms of cost, maintenance and simulator image quality.

Projection systems, as the name implies, project the generated image either directly from the rear or to the front of the screen. Rear projection images are necessarily attenuated in effective resolution and display luminance while front projection systems affect the amount of contrast available. In some cases, projection systems may use mirrors which, in turn, reflect the image on the screen and these mirrors may add varying amounts of optical distortion to the displayed image. Dome projection systems, which project the image on to a very large curved surface, are particularly prone to image distortion problems.

Projection systems are primarily used because they allow for image displays over a very wide area. Display systems with such very large FOVs are difficult to achieve or are cost-prohibitive with other forms of display technology. For example, the projected image of an addressable resolution of 1280 x 1024 can be spread over a screen area covering several meters vertically and horizontally. However, by projecting an image over wide areas without increasing the basic addressable resolution means that the *spatial* resolution will be dramatically reduced when compared to the same addressable resolution displayed over a smaller image display field. The poorer spatial resolution may translate into an effective resolution which is unacceptably low for the desired application. If the lowered spatial resolution is acceptable, projection display systems are a very cost-effective solution to large FOV design requirement.

Another potential disadvantage to projection systems is that the luminance levels of the displayed image tend to be much lower than that of other display systems. Projection light sources are limited in the amount of light that can be emitted by the projector lamp. This limited light energy needs to be distributed over a very large area so that the average luminance of the display declines, sometimes dramatically. Screen reflectance or translucence values and other characteristics may attenuate the luminance value of the display even further. While often applied to fixed-base flight simulators, projection systems are generally not suitable for simulators equipped with motion platforms. The large display surface which is characteristic of projection systems typically requires that they be fixed to the floor of a facility and detached from the simulator cab.

Direct View Displays

An alternative to the projection system, which is in use in many flight simulators, is the *direct view* system. The direct view system provides the pilot with a direct view of the simulated visual scene on a Cathode Ray Tube (CRT) or other form of electronic display system, such as a Liquid Crystal Display (LCD) or a gas plasma display. The main advantage of direct view is simplicity since no complex projection, beam-splitter or other systems are required to display the CGI image. Furthermore, as the direct view systems are mounted directly on the simulator, the use of motion platform is not a problem. Additionally, the direct view system can deliver much higher luminance levels than projection systems and can readily meet the 300 cd/m² highlight luminance levels recommended in this book. Furthermore, because the same addressable resolution is now displayed over a much smaller area, a much higher spatial and effective resolution is available with direct view systems than is generally possible with projection systems. The main disadvantage of the direct view system is that, in order to meet the same FOV requirements attained by projection systems, a combination of a large number of individual direct view displays may be needed. This adds significantly to the display system cost and complexity of such systems when large FOV displays are desired.

One very important disadvantage to direct view systems is that, when the viewing distance of the display is less than 1 m from the design eyepoint, the direct view image display may significantly affect a pilot's perception of object distances in the simulated scene. When direct view displays are viewed at these short distances, the pilot's eye will accommodate (focus) to an object at the distance of the display and not accommodate for the actual viewing distance at which the object is seen in the real world. The same holds for the binocular convergence of both eyes - convergence will be greater than if the object were viewed at the same distance in the real world since the eyes are converging on the object at the distance of the visual display not at the intended, simulated distance of the object. The consequences for flight simulator display design are not trivial. A recent study by Pierce and Geri (1998) found that an object displayed on a direct view system viewed at distances less than 1 m appeared 15-30 percent smaller than when the same object was presented at the optically correct distance from the viewer. This underestimation of the size of objects in the direct view display may lead to pilot misperceptions of the distance of objects displayed in such systems. Misperceptions could include runway width and length as well as the perceived distance of airborne or ground-based objects that might be displayed.

Direct View/Collimated Displays

One obvious solution to the problem of accommodation-vergence in direct view systems is to place the direct view displays at least 1 m or more from the pilot's eyepoint on all simulators with this type of display system. However, increasing the distance of the display from the pilot's eyepoint reduces the effective FOV of the display. This loss of effective FOV would have to be compensated for by increasing the size of the display monitor or by adding additional monitors to the system. Fortunately, there is a less expensive alternative. By placing a collimating lens between the monitor and the pilot, it is possible to render the displayed image at optical infinity. This process, termed *collimation,* is illustrated in Figure 1.7.

Collimated direct view display systems, or simply *collimated displays,* eliminate the accommodation-vergence problem of direct view displays by focusing the displayed image at optical infinity. The pilot's eyes are now focused at a distance which is optically beyond the range at which accommodation or convergence of the eyes will be affected. There are also additional benefits afforded by collimation. Many pilots will experience a greater sense of apparent depth or distance in the displayed image that is absent from direct view imagery. Most advanced simulators now use some form of collimated display system if a direct view system is employed.

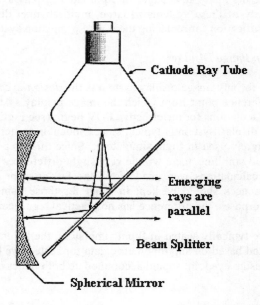

Figure 1.7 Collimated Visual Display System with Beam Splitter
(Image Courtesy of NASA)

There are some disadvantages to collimated display systems, however. The introduction of the collimating lens typically attenuates the luminance available from the display monitor. The effects of collimation on display luminance need to be factored into the simulator specification during design. Another disadvantage with collimated displays is that more space is required to accommodate them. This is due to the need to allow some distance between the collimating lens and the display monitor in order to produce a correctly focused image.

To accommodate collimated direct view systems, more compact image display designs have been developed. Figure 1.7 also shows the use of beam-splitters, which make the size of collimated, direct view systems more compact. Such a configuration is often used where it is necessary to provide different viewing channels to a single pilot, such as a separate view through a side cockpit window. The beam-splitters serve to redirect the displayed image from the monitor in order to present an image that is correct for that particular area of the cockpit canopy or windscreen.

The typical cockpit visual display system configuration juxtaposes several collimated direct view systems to cover the pilot's frontal view as well as the near and far visual periphery. Combining multiple viewing channels requires highly accurate merging and continuity of adjacent image displays to avoid distracting image artifacts. This adds to the complexity of both imaging software and display hardware. Despite the many advantages of collimated display systems, the added cost and complexity of these systems in large, multi-channel display systems is often used as a justification for replacing them with projection systems.

Design Eyepoint in Image Displays

The starting point for any image display system is the *design eyepoint*. The design eyepoint is the reference point from which the image display system is developed and from which calculations for the effective FOV begin (see Figure 1.8). A single eyepoint for each display system is typical since binocular or stereoscopic display systems are very rarely used in flight simulators. Since this eyepoint is calculated from the pilot's body midline, there will be very slight differences in image display geometry than if calculations were made from the left or right eye of the pilot. While these differences should be kept in mind for those using simulators for display research purposes, the difference has no practical significance for simulator training.

Since pilots are typically seated in flight simulators, the height of the eyepoint should be calculated based on anthropometric data for seated eye height in a given population. The design eyepoint should accommodate both males and females.

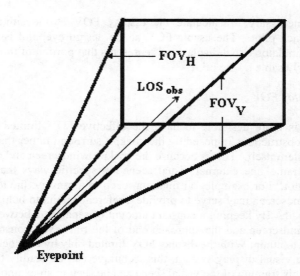

Figure 1.8   Design Eyepoint, Field-of-View and Observer Line-of-Sight

The distance of the design eyepoint from the image display systems is calculated from the frontal plane of the display surface to the pilot's eye. This distance will vary somewhat from one pilot to the next due to differences in pilot stature, as pilots need to move the seat fore and aft or up and down in order to reach the rudder pedals and other controls.

The design eyepoint calculations will also depend upon the type of display system used. Projection systems will typically use a single design eyepoint calculated from a single point between the two pilots. Direct view and collimated direct view display systems may use two separate design eyepoints, one for each pilot position. This is due to the fact that the use of collimating and beam-splitter lenses limits the optimal viewable area of a given display. This display 'exit pupil' means that only one pilot will be able to view the displayed image for his or her position. For example, an object viewed in a side cockpit window display of a pilot in the left seat of the simulator will not be viewable by the pilot in the right side. This multi-view design assures that each pilot will be presented with an accurate simulation of an external visual scene, but does so by restricting each pilot view to only a single pilot position.

Display Field-of-View

The FOV of a simulator visual display is determined by the total amount of visual scene display required to accomplish the purpose for which the simulator is built. This may result in a FOV which is much less than that actually available in the aircraft. Additionally, the usable portions of the display FOV may be affected by display optics, simulator components, cockpit structures and other factors. For this reason, it is necessary to calculate the display FOV at the simulator design eyepoint for each pilot. The usable FOV at the design eyepoint is the *effective* FOV for the simulator visual display and represents that portion of the visual scene display actually usable by the pilot.

Effects of Display FOV

In general, it is more desirable to have the effective FOV limited only by the occlusions or obstructions imposed by the cockpit surround rather than limited by the display system itself. This is because the cockpit windscreen and other cockpit components 'frame' the external visual scene[5] in specific ways that may be of value to the pilot. For example, on final approach to a runway, the lower edge of the cockpit windscreen may serve to provide a reference point to help establish the correct glide path. By keeping a constant amount of separation between the lower edge of the windscreen and the approach end of the runway, a constant approach angle is easier to attain. With the display FOV limited only by the cockpit structure rather than the visual display system, this technique can be readily transferred to the aircraft. This framing of the visual scene can be seen in an example of a visual display system in a modern flight simulator (Figure 1.9).

Figure 1.9 Visual Display System in B-747 Flight Simulator
(Image Courtesy of NASA)

The relationship between cues provided by the cockpit surround and external objects in the visual scene are also found in formation flight. In this case, cockpit canopy components and components of the aircraft can be aligned to form a correct 'sight picture' when the two aircraft are in the correct position. A similar phenomenon occurs in aerial refueling (Lee and Lidderdale, 1983). When the

---

[5] Visual scene as defined here refers to the scene that appears to the pilot out of the cockpit windows in operational flight.

simulated aircraft is close to airborne or ground-based objects, the sight picture formed by the cockpit windscreen and the elements of the external visual scene provides important visual cues to the pilot.

It is obvious that the effective FOV needs to be large enough to view certain objects to support the training, evaluation or investigation of certain flight tasks. For example, the runway needs to be seen on final approach at a reasonable distance if the simulator is to support the approach and landing task. Similarly, the display of other aircraft is needed in various areas of the display FOV if realistic collision avoidance or air combat training is to be carried out.

The need for a larger FOV in a flight simulator is less obvious when there is no clear need to display objects in the area of the visual scene provided by the added FOV. Indeed, many flight simulators have image display systems with very limited FOVs for precisely this reason. Many smaller training devices targeted for instrument training, for example, provide single channel image display FOVs of 48 deg (h) x 36 deg (v).[6] Such limited FOVs are usually justified by the limited training objectives of these devices.

If the objective is control of the aircraft by means of reference to the visual scene then the requirements for FOV, particularly FOV_H, become more stringent. When flight control is dependent on information from the external visual scene, the display FOV_H provides important cues for aircraft attitude control, particularly in pitch and roll axes. Adequate FOV in a visual display system intended to provide the pilot with these cues needs to address both FOV_V as well as FOV_H. In the normal, non-aerobatic flight regimes, pitch axes cues from the visual scene visible horizon will be assured if the FOV_V display is the same as the FOV_V of the windscreen of the simulated aircraft.

Adequate FOV_H is also needed to provide the pilot visual information in two important areas. The first is control of the aircraft's roll axis. While most flight simulator visual FOV_Hs range between 30 and 75 deg horizontal, a recent study with artificial horizons suggest that this limited FOV_H may not be enough to present all of the roll cues normally provided to the pilot by a real world visual horizon (Comstock, Jones, and Pope, 2003). In this study, improvements in pilot roll control were seen with increases in FOV_H up to and including 110 deg. The implications of this study is that effective FOV display systems measuring at least 110 deg horizontal may be needed if the flight simulator visual display is to provide the full roll control cues that can be used by the pilot.[7]

A very wide FOV_H in a flight simulator visual display also provides an increased likelihood of inducing self-motion or *vection*. Vection is the sense of self-motion that is created by exposure to moving visual fields. As with roll motion cues, the strongest vection stimuli are visual scene movements or optical flows in the visual periphery with eccentricities beyond 60 deg (Brandt, Dichgans,

---

[6] It is not by accident that such systems reflect the 4:3 aspect ratio of a conventional CRT monitor since that is what is often used as a display system.

[7] Note that this is substantially greater than the 75 deg FOV_H currently required by the Federal Aviation Administration for Full Flight Simulation visual display systems.

and Koenig, 1973). Wide FOV_H in flight simulators provides the opportunity to induce vection by presenting visual movement in the pilot's far visual periphery. Inducing vection, however, also depends on the presence of scene texture or object detail necessary for the creation of an optical flow field. This typically occurs close to surface terrain so vection in the simulator is likely to be created only during takeoffs, landings, taxiing or in very low level flight operations. While wide FOV_H allows for the possibility of illusory self-motion along the aircraft's longitudinal axis (*linear vection*), they can also produce *circular vection*. Circular vection is illusory, rotational self-motion that might occur when an aircraft is rolled about its longitudinal axis. In both forms of vection, the illusion of self-motion may be very powerful. As self-motion implies aircraft motion to the simulator pilot, vection plays an important role in the illusion of flight. However, a very wide FOV_H display system also has the potential to create significant problems for the simulator pilot that would not normally occur in real flight (see Ch. 8).

Summary

Visual scene simulation in flight simulators has evolved from simple line drawings projected on a screen to complex, photorealistic images. Both image generation and image display design characteristics of a simulator's visual scene simulation have a significant impact on what the pilot will be able to see. The requirements for both image generation and display need to be driven by the capabilities of the pilot's visual system to process information as well as by the tasks that the pilot needs to perform. Simulators designed to provide the necessary visual information for the training and evaluation of visual flight tasks must have visual scene simulation capabilities which conform to much higher performance standards than training devices designed for other purposes. Recent advances in visual scene simulation now make possible the widespread use of simulators for the training of even the most basic visual flight tasks.

Chapter 2

Sound Effects and Communications Simulation

Introduction

While the simulation of visual scenes has been significantly improved over that available only a few years ago, the ability to create a virtual environment of sound within flight simulators has also improved substantially. However, sound simulation is often overlooked when flight simulator specifications are considered. This is partly due to the fact that sound is a more subtle aspect of the cockpit environment, more likely to attract attention by its absence than by its presence. It is also partly due to the fact that sound and communications simulation are elements of the flight environment which are usually considered to contribute much less to simulator realism than other simulation elements, such as visual scene and motion simulation. For these reasons, simulation research and development efforts have focused more on the latter aspects in the design of flight simulators than on sound simulation. Nonetheless, sound and communications simulation play an important role in creating a realistic flight simulation environment and are deserving of attention.

Sound Effects in the Cockpit Environment

Sound plays a complex and varied role in the cockpit environment. Sound serves to provide the pilot with feedback on the operating status of aircraft engines, for example, feedback which is critical to safe flight operations. Not only does the amplitude and frequency of engine noise provide cues to their normal operation, but changes in engine sound patterns can be early indicators of engine malfunctions. Additional sound effects are associated with wind, rain and hail striking the windscreen and other cockpit structures. These sound effects are also correlated with certain aircraft state parameters, such as airspeed as well as being useful indicators of the severity of weather encountered by an aircraft.

Operation of aircraft control surfaces and landing gear may produce unique noise patterns from the workings of gears, electric motors and other machinery. Control surfaces and gear extension at too high an airspeed can cause distinct changes in wind noise heard within the cockpit as well as significant cockpit vibrations. Distinctive sounds are also associated with tire braking and skidding on the runway or taxiway surface.

Sound effects serve other purposes as well. Sound is a very effective alerting stimulus because, unlike visual stimuli, it does not require an orienting response from the pilot to be effective. This allows the pilot to detect changes in engine sounds patterns, for example, before they might be seen on engine instrumentation.

Sound effects simulation may play its most important role in creating the sensation that the simulated aircraft environment is behaving in the same predictable way that one would expect in the real environment. Sound serves in this sense as confirmatory information on the state of the environment as well as an alerting stimulus. It is perhaps this fulfillment of the *expectations* of the pilot associated with this confirmatory evidence that makes realistic sound effects simulation worth the investment.

Avionics

Sound simulations of aircraft alerting and warning avionics systems are even more important. Accurate replication of alerting systems such as stall and gear unsafe warning horns and engine fire alarms are necessary to assure that pilots learn to recognize their unique sound signatures. Realistic simulation is also necessary for those systems that incorporate voice display alerting and information such as Ground Proximity and Warning Systems (GPWS) and Traffic Collision Avoidance Systems (TCAS). Such systems contain voice displays that need to be distinguished from other human voices that may occur in the cockpit. The simulation of these voice displays needs to be highly accurate in order to faithfully reproduce the complete sound signature of the system involved. When the pilot transfers from the simulator to the aircraft, there should be no notable differences between simulator and aircraft avionics voice displays.

Sound simulations of navigation beacon signals are also needed for navigation avionics. Most important of these are the aural tones associated with Instrument Landing System (ILS) marker beacons and the aural Morse Code Identifier signals from radio navigation beacons. As with avionics display sound simulation, the sound signature of these navigation aural displays needs to be simulated accurately, particularly with regard to the frequency and inter-signal interval of the tones provided.

Sound Effects and Human Audition

While visual scene simulation has seen dramatic improvements in the last two decades, the digital revolution has also had an impact on the creation of virtual sound environments. Realistic sound effects simulation, like visual scene and whole body motion simulation, requires an understanding of both the pilot's capability to make use of sound effects simulation and the technology by which it is produced.

In flight simulators, sound effects are typically generated by replaying pre-recorded sound effects files stored digitally in system memory. When a simulator event is encountered which requires a sound effect, the appropriate file is retrieved from the system and played back through the simulator speaker system. In general, these sound effect files are created from actual sound recordings taken from aircraft cockpits. In other cases, the files are created using a combination of effects taken from a variety of sources and then edited to create the desired sound effect. In some cases, such as wind or rain noise, effects files are played back repetitively as long as these conditions are encountered, while in other cases, the effects respond directly to aircraft control inputs. In the latter case, for example, changes in the amplitude and frequency of engine sound effects are produced as a direct function of pilot throttle inputs.

Humans are limited in their ability to process sound effects to frequencies within 20 Hz to 20 kHz. For older adults, particularly males, hearing loss for frequencies above 15 kHz is not uncommon. Obviously, reproduction of sound effects in a flight simulator outside of the normal range of human hearing serves no useful purpose. For example, certain engine sounds and other sound effects of aircraft operations will fall below the 20 Hz lower limit of human audition and need not be simulated.

The ability to hear sounds is affected by both the frequency and the loudness of the sound. Higher frequencies of sound need to be presented at higher decibel levels in order to be audible. Sound effects that contain high frequency components of, for example, 5 kHz or higher, may need to be presented at decibel levels above that of human speech to be audible and this may interfere with the audibility of speech between pilots or between pilot and instructor. Normal speech occurs between 40 and 60 dB of volume at frequencies between 500 Hz and 5 kHz. A decision has to be made by the designer of the simulation as to whether or not the potential for interference of sound effects in these high frequency ranges justifies its simulation.

Humans are quite adept at discriminating signature patterns so that differences as small as 2 to 5 Hz can be identified. This is particularly important for the simulation of sound effects frequency spectra and for the reproduction of the auditory display component of alerting and warning systems. Accurate reproduction of the sound signature of the effect or system is essential. Learning to recognize these sound signatures and transferring that recognition ability to the aircraft is essential for effective simulation training.

Sound Displays

The display of sound effects needs to match the pilot's capabilities at detecting and discriminating sound. This means that the sound displays or speakers have an amplitude and frequency response comparable to that of the pilot. Additionally, distortions to the playback of the recorded sound effect are all too common in flight simulators due to the poor placement within the simulated cockpit. Speaker placement can significantly distort the actual sound generated owing to the acoustic properties of the surrounding space.

Placement of speakers in the simulator is also important due to the component of human audition called *localization*. Because humans have two ears located on opposite sides of the head, the difference in the time it takes for a sound to reach the two ears is used to identify the location of the source of that sound. For example, if the sound reaches the right ear slightly before the left, the sound source will be identified as located in the right hemisphere. Simulation sound effects designers need to understand this characteristic of human audition in order to correctly manipulate the pilot's perception of the location of the sound. This is particularly important in the simulation of sounds associated with engine failure in multi-engine aircraft. In this case, the accurate simulation is needed not only of the correct sound frequency, but the correct localization of the sound as well.

With the advent of digital signal processing and computer controlled signal display systems, it is possible to provide the full range of cockpit sound effects with a relatively inexpensive array of speakers. A recommendation for such a system is to provide four speakers, two in front and two behind the pilot (Shilling and Shinn-Cunningham, 2002). The front speakers should be placed at ± 30 deg in front of the pilot and the rear speakers ± 110 deg behind the pilot. This arrangement, in conjunction with digital signal processing appropriate to the sound effect, can provide the full range of localization and surround effects desired in the flight simulator.

In creating a simulation environment for the pilot or aircrew, realistic re-creation of sound effects associated with normal and abnormal aircraft operations are essential, as is the accurate reproduction of sounds and voice displays integral to alerting and warning systems. Beyond the reproduction of these sound effects and displays is perhaps an even more important simulation requirement—that of radio communications simulation.

Radio Communications

One of the more difficult tasks for flight simulation design has been the provision of realistic radio communications simulation. Radio communications are a vital element in all aircraft operations, civilian and military. For flight simulators, two major areas of communications simulation are of great importance in creating a realistic flight environment. The first is broadcast communications, i.e., those pre-recorded communications transmitted on a radio frequency which are intended to be of general interest to all aircraft on the frequency. The second category of radio communications are Air Traffic Control (ATC) communications. ATC communications are specifically intended for aircraft under the control of the ATC facility served by the radio frequency. These communications serve a variety of purposes, but the central purpose is to provide controlling information to those aircraft within the facility's airspace. Both categories of radio communications, broadcast and ATC communications, are an integral part of the flight environment.

Broadcast Communications

Automated weather and terminal information broadcasts are the simplest forms of radio communications. In the US, these include the Automated Surface Observation Systems (ASOS), the Automated Weather Observation System (AWOS), and the Automated Terminal Information System (ATIS). These systems are designed to provide weather and other airport area information to the pilot in order to aid in planning departures and arrivals. The broadcasts are updated at least once per hour and are intended to provide the pilot with current information on airport area conditions. Some of the systems still use manual analog recordings while newer systems use more advanced text-to-speech (TTS) systems for broadcast transmissions. The latter are designed to complement automated observation systems such as ASOS. Regardless of the system used, the pilot will expect a close correspondence between the broadcast conditions and actual conditions. This is especially true of weather information such as ceiling, visibility, wind speed and direction.

ATIS and ASOS systems are the most widely used for terminal area information in the US. They are broadcast over reserved radio frequencies within a given airport terminal area so they will not interfere with other radio communications traffic. Most aircraft are equipped with at least two radios capable of receiving VHF signals, so it is not uncommon for a pilot to monitor these broadcasts over one frequency while using a second frequency for ATC communications.

Some en route weather broadcast information may also be transmitted from navigation radio beacons. The broadcast is typically transmitted over a navigation radio frequency and can be received only when the aircraft is within reception range of the transmitter. In the US, one such system, the Terminal Weather Enroute Broadcast (TWEB) provides important information on en route weather including significant meteorological events such as thunderstorms, icing and turbulence.

The most important characteristics of these broadcast systems for the pilot is the currency of the information contained in the broadcast, the type of information available, and the accessibility of this information during flight operations. The information in the broadcast may significantly impact pilot decision-making regarding route planning, fuel consumption, diversionary airports and other flight critical issues. For simulator training, evaluation or research into pilot decision-making and planning behavior, the provision of these broadcasting systems in the flight simulator is essential.

Air Traffic Control Communications

ATC Communications represent the most important use of radio communications in aviation operations. At its simplest level, ATC communications consist of a broadcast by the pilot of his or her position or intentions over a reserved frequency. In the US, these Common Traffic Area Frequencies (CTAFs) serve airport terminal areas without operational ATC control towers. The broadcasts are transmitted on a very limited number of reserved radio frequencies. Because of this limit in the number of reserved frequencies, it is not uncommon for transmissions intended for one airport traffic area to be heard by pilots in several others. To avoid any confusion that may result, pilots using the CTAF must begin and end radio transmissions with the name of the airport traffic area for which the transmission is intended. As there is no control tower or radar coverage for these uncontrolled airports, information transmitted on a CTAF is essential in maintaining safe air traffic operations.

Most flight operations, including most commercial and military flights, operate under some form of air traffic control system. This control system requires that two-way radio communications be maintained at all times between the aircraft and the controlling facility. In civilian airports, control is maintained over ground operations and runway operations by separate ground and tower controller facilities. Shortly after takeoff, the aircraft may be transferred to a departure controller who may subsequently turn the aircraft over to other departure controllers or to a center control. As the flight proceeds, it is typically transferred from a center controller to departure control and thence to tower and ground controllers. Since the aircraft is under continuous control from an ATC facility on the ground, the pilot must maintain and monitor communications with that authority at all times during the flight.

Ownship and Frequency Chatter Communications

There are two classes of communications in ATC that are important to the pilot. *Ownship communications* are those communications from ATC directed specifically to the pilot's aircraft. All of these communications must contain the aircraft's call sign when transmitted to and from the aircraft. There are other communications that are transmitted on the same radio frequency as ownship communications, but these are transmissions between other aircraft and the facility on that frequency. These latter transmissions are termed collectively as radio frequency chatter or simply *frequency chatter*. Since several aircraft may be sharing a given frequency, frequency chatter communications usually comprise the bulk of the radio communications that a pilot will hear. Generally, the pilot's primary task is to monitor the radio frequency with enough attention so as to assure that communications meant for his or her aircraft are not missed. At the same time, monitoring chatter on the frequency can provide the pilot with useful information such as the location of other aircraft, weather encountered by aircraft on the same route, and potential delays at the destination airport.

ATC communications are, of course, two-way communications. Pilots may initiate the communication with the controller or the communication may begin with a transmission from the controller. In either case, only one person on a given frequency can speak at any one time. This *asynchronous* radio communication system can, and does, lead to delays and other anomalies in communication between pilots and controllers. In busy airport terminal areas, delays in communications are not uncommon due to the near continuous stream of chatter filling the radio frequencies.

Given the prevalence of radio communications in the operational environment, one would expect that modern flight simulators would provide communications simulation routinely. Realistic communications simulation would certainly be expected for what are termed Full Flight Simulators (FFS) used by commercial air carriers for Line-Oriented Simulations (LOS) where realistic radio communications simulations are paramount. However, a survey of US commercial airlines indicated that communications simulation is provided not by the flight simulator, but by the instructors themselves (Burke-Cohen, Kendra, Kanki, and Lee, 2000). In the airline training simulators represented in the survey, the instructor role-played the ATC controller as well as broadcast transmissions such as ATIS. Instructors are, of course, not air traffic controllers so this type of ATC communications simulation is predictably low in realism in several respects. Apart from simplified ownship communications, instructors normally are not able to provide frequency chatter. As instructors must provide ATC communications simulation in addition to their primary teaching and evaluation duties, the quantity and quality of communications simulation generally suffers.

The low level of radio communications simulation fidelity in simulation training environments needs some explanation. In fact, there are at least two reasons why this area has not received greater attention in flight simulator development. First, neither the FAA nor any other regulatory agency requires that simulators provide realistic communications simulation in order to train or evaluate pilots. Without the regulatory compliance requirement that exists for other simulator components, such as motion cueing, there is little incentive to develop communications simulation technology for flight simulators.

Additionally, the lack of communications simulation realism in the modern flight simulator is at least partly attributable to the lack of evidence that a high degree of realism would contribute significantly to training and evaluation of pilots. In short, some in the training community may believe improving communications simulation may add to the realism of flight, but that realistic communications simulation contributes little or nothing to the instructional value of flight simulator training. In the absence of any evidence to the contrary, this is not an unreasonable assumption. A recent series of studies comparing high and low communications simulation realism, however, suggests otherwise (Lee, 2003). In these studies, aircrews that were provided with realistic ownship communications and frequency chatter performed very differently from those crews provided with only the instructor playing the role of controller. Crews initiated more ATC communications with high communications simulation realism and generally reported higher levels of mental demand and effort. A second study showed that pilot workload was measurably higher even when only realistic radio frequency chatter was simulated. The implications of these findings is that the existing methods of communications simulation are not adequate and may be providing simulation environments where aircrew communications behavior and levels of mental workload are not comparable to that of the operational environment.

Along with this evidence for the training value of realistic ATC communications simulation is the additional finding from the Lee (2003) and Burke-Cohen, et al (2000) studies that both flight instructors and pilots believe that more realistic ATC communications simulation is needed for effective simulator training. With this demonstrated need for improved ATC communications simulation, a sustained and focused effort to enhance enabling technologies is also needed.

Radio Communications Simulation Technology

There are two basic components required for realistic communications simulation: 1) the production and display of intelligible and naturalistic speech and 2) highly reliable, speaker-independent recognition of continuous speech. In the display of speech, analog recordings of human speakers would seem a logical choice, and in fact were used in early simulators to provide highly realistic radio broadcast simulations. Later improvements in analog displays digitized individual words and phrases and played them back in the designated order and the appropriate time.

While analog displays provide a high degree of realism, they have a number of serious shortcomings. Analog displays are dependent on a unique source for recordings. Since humans are highly adept at discriminating among human voice patterns, analog displays cannot be modified by anyone other than the original speaker. With the many voices required for communications simulation, this problem is multiplied many times. Furthermore, the intonation patterns of human speech make rearranging words and phrases to meet new communications simulation demands extremely difficult.

Text to Speech

Fortunately, there have been significant developments in alternatives to analog speech displays that make them suitable for communications simulation. In many cases, the new synthetic speech displays provide speech simulation quality indistinguishable from analog recordings. The synthetic speech displays are based on text-to-text speech (TTS) technology. As its name implies, TTS technology converts the electronic text into its speech equivalent. This text can occur in a variety of storage media or within a computer program. The conversion is done with a TTS engine which examines individual words, identifies them and selects a pronunciation appropriate to the word and the linguistic context within which the word is found. TTS engines exist for a large variety of languages. Generally, such engines will attempt to provide the pronunciation appropriate for the language so such systems are not normally designed to display speech in English accented in French, German or some other language. This a feature which would be potentially useful for exposing pilots to communications from non-native, English-speaking controllers.

Two general types of TTS engines are typically used; one based on a physical model of the human speech-production apparatus and one based on concatenation of individual speech sounds (diphones and triphones). Model-based TTS engines have the advantage of being able to simulate a virtually unlimited range of human languages and dialects. Such systems have been able to produce intelligible speech, but have more difficulty delivering natural-sounding speech. Engines based on diphone or triphone concatenation, in contrast, tend to produce speech that is more natural in sound than that produced by model-based TTS engines. However, the concatenation engines are based on actual samples of human speech patterns. The process is lengthy and complex and inherently less flexible than model-based engine technologies since it is dependent on sampled human speech rather than a general model of the human speech system.

Both methods are capable of providing a high degree of intelligibility in the speech produced. *Intelligibility* refers to the ability of the listener to correctly recognize words produced by the TTS engine. The vocabulary for standard phraseology in aviation is relatively small compared to most natural languages. As a consequence, highly intelligible TTS speech displays may be easier to produce for communications simulation than for other applications where the vocabulary size is much larger. In any case, intelligible TTS displays are no longer a technical challenge and should see use in communications simulation systems in flight simulators just as they have in other aviation applications such as automated weather broadcasts.

The naturalness or 'human-like' quality of TTS speech displays, however, can vary widely and has generally lagged behind intelligibility in the development of TTS displays. The naturalness of TTS speech displays is affected by the inability of many TTS engines to accurately reproduce the prosodics that are so integral to normal human speech patterns. *Prosodics* include the pacing, intonation and stress patterns of human speech. The absence of these patterns results in a mechanical or unnatural speech. Naturalness can also be lost when otherwise natural-sounding TTS speech systems are required to deliver speech at much higher rates than normal. For the ATC component of communications simulation, text-to-speech engines will need to deliver speech at a rate matching those of controllers. These speech rates typically average 50 to 100 percent higher than normal conversational speech rates (Morrow, Lee, and Rodvold, 1993).

Naturalness in TTS speech displays used for radio communications simulation is important to avoid the presence of distracting speech artifacts. In components of communications simulation which are intended to simulate normal human speech, the speech display should be indistinguishable from that which could be provided by an analog recording of a human speaking the same text. There are, of course, exceptions to this rule. Certain speech displays associated with alerting and warning systems have incorporated the 'unnatural' sounding TTS speech displays precisely *because* they attract attention. These speech displays should be reproduced exactly in the simulator in order to expose the pilot to identical speech displays in both the simulator and the aircraft.

In order to provide the necessary level of realism in communications simulation, TTS displays need to provide a large of unique voices. Unique voices are needed for terminal and en route broadcasts, for individual ATC facilities and sectors within facilities, and for varying the number of pilots in radio frequency chatter. In selecting voices for communications simulation, designers should avoid using the same voice for ATC ownship communications in adjacent facilities or sectors since voice changes provide feedback to the pilot that he or she is in fact in communication with a new facility or sector. Confusion can also be avoided by not using controller voices to simulate weather broadcast or radio frequency chatter. For the typical LOS simulator flight which lasts from one to two hrs, 10 to 12 unique voices will probably be needed for realistic communications simulation. Shorter simulated flight scenarios, of course, will need fewer unique voices.

Communications Simulation Content

A recurring problem with the few communications simulation systems that have been deployed in flight simulators has been with their content. For example, ATIS messages with incorrect weather information, frequency chatter using airline call signs that no longer operate on the route, and the use of radio frequencies that do not correspond to approach plates and charts. These problems reflect poor systems integration and a failure to maintain current information in the radio communications simulation database.

Systems integration of communications simulation is nearly as important as the communications simulation technology itself. For an ATIS or other simulated terminal broadcast to be useful, the simulator needs to know details about current weather, runways in use and other airport-related issues that a pilot would expect to hear in such a broadcast. This can only be done if the communications simulation component of the simulator is continuously updated with information from other components of the simulated airport environment. A simple pre-recorded message will not be useful if the simulator scenario conditions change over time, as is often the case.

Communications content are subject to change for other reasons as well. As the real aviation and airspace operating environments are subject to change, the communications simulation systems must regularly be updated to reflect changes to that operating environment. These include changes to communication and navigation radio frequencies, the addition and deletion of aircraft call signs, and changes to traffic routing and altitude assignments. Communications simulation systems which are developed for flight simulators must support a means by which these changes can be made in a timely and cost-effective manner. Timely updates to the content of a communications simulation database are needed to avoid presenting information from the simulator which may contradict other information available to the pilot from charts, approach plates, and other sources.

Speech Recognition and ATC Communications

Perhaps the most daunting of technical challenges in flight simulation is the provision of a fully automated ATC simulation where there is not only the realistic display of radio communications messages, but there is also the ability of the simulated air traffic controller to understand and respond to pilot communications. Such a capability is an essential requirement for a fully integrated ATC communication system in flight simulators, but it relies on a component technology that has been slow to develop.

This component technology is speaker-independent, continuous speech recognition. Such systems do not require speaker training and can recognize speech delivered at the normal rate without artificial pauses between words. This technology differs significantly from command speech recognition systems in which only single words or phrases are recognized. Historically, continuous recognition systems have had difficulty delivering high rates of word recognition sufficient to permit their use in flight simulators. Recognition rates comparable to that of normal human speech, which is above 99 percent, are needed to avoid introducing communication artifacts into the simulation environment. Recurring problems with ATC communications speech recognition would not only introduce additional pilot workload, but reduce confidence in the simulator's overall reliability as a training device.

Recent developments in speech recognition engine technology suggest that deployment of a fully automated ATC communications system is possible in the near future but may occur with some restrictions. Field deployments of recognition systems in ATC tower simulator training simulators have now been carried out successfully (Tomlinson, 2004). These systems rely on a combination of artificial intelligence and the limited operational vocabulary of ATC standard phraseology to produce high speech recognition rates. It is possible that such a system could be used in flight simulators provided a speech recognition vocabulary restricted to standard phraseology were acceptable to instructors and pilot trainees. Such systems would preclude the use of non-standard phraseology by pilots—a restriction that may be resisted as too rigid by some while considered a useful means of enforcing standard communications phraseology by others. Notably, large deviations from non-standard phraseology typically occur when unusual events or anomalies occur, as often happens in training scenarios. These are situations for which current speech recognition systems are not well suited. Such systems do have the capability of relaxing the strict phraseology requirement so that a certain degree of deviation from standard communications phraseology is tolerable. This may come with some loss of system recognition and response performance, however. Making use of current, advanced speech recognition systems in flight simulators will probably require tradeoffs between the desire for high reliability on the one hand, and the desire for flexibility in communications phraseology on the other.

High speech recognition reliability is essential for development of a fully automated ATC communications simulation capability. If such systems could be deployed, normal radio communications protocols including clearance requests, ATC communications readbacks, facility check-ins, diversion requests and re-routings, traffic alerts and other communications and queries would be possible. Such a capability would allow much higher levels of flight simulator realism and reduce the associated workload that the task of ATC simulation currently imposes on instructors. Such systems would also allow better use of simulators for skills maintenance without the presence of instructors. Recurrent instrument proficiency maintenance, for example, would be significantly enhanced by realistic ATC communications simulation which is not currently available in simulators used for this purpose.

Advanced speech recognition technology combined with advanced instructional technologies also has the potential for more efficient and more effective use of flight simulation technology in general, including the development of intelligent instructional systems which could provide simulator skills training and maintenance without the presence of a flight instructor (see Ch. 9).

Summary

Sound effects and communications simulation are an integral part of any flight simulator. Advances in sound generation and display technology now allow highly realistic sound simulation capability at a relatively low cost. Broadcast and ATC communications simulation capabilities are also essential for effective training in flight simulators, particularly those used for line-oriented training purposes. Advances in speech recognition and speech display technologies have now reached the stage where a fully automated communications simulation system for flight simulators is now possible.

Chapter 3

Whole Body Motion

Aircraft Motion

All aircraft move in a three-dimensional space. As the aircraft moves it necessarily imparts forces on the pilot's body. These whole body motion forces are sensed by the pilot through a variety of non-visual systems. The forces imposed on the pilot by an aircraft and the pilot's response to these non-visual motion sensations play an important role in a variety of flying tasks. Attempts to recreate these sensations in ground-based simulators have a long, and somewhat controversial, history in flight simulation.

In flight operations, an aircraft may impart sensations to the pilot's body in the form of translational and rotational forces in any one or more of three aircraft axes. These are the longitudinal axis (x), the vertical (z) axis and lateral (y) axes (see Figure 3.1). Two different forms of motion, translation and rotation, can be applied to the three aircraft axes to produce a total of six degrees-of-freedom (6-DOF) of motion. In the aircraft, these motion forces occur with a wide range of amplitudes and frequencies and may have different sources and different consequences for pilot control.

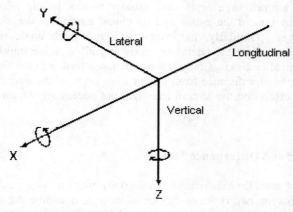

Figure 3.1 Aircraft Axes of Motion

The translational motion forces affecting the pilot are most commonly referred to as positive or negative gravitational (G) loads. In non-aerobatic, fixed-wing, aircraft maneuvering, these loads tend to be relatively small and infrequent. In the longitudinal axis of an aircraft, translational motion occurs most frequently during accelerations in takeoff and decelerations in landing. Translational motion in the vertical (z) axis occurs during the initial stages of takeoff and landing, during steep turns, and as a result of encounters with turbulence. Although translational motions in the *y* axis are less common than in other aircraft axes, these forces do occur occasionally, most often as a result of disturbances to the aircraft from the environment or from aircraft system failures. Pilots will generally experience only positive G-loads, though occasional negative G-loads may be experienced, particularly in aerobatic maneuvers. In aerobatic environments, such as air combat maneuvering, translational forces are much more intense, occur more frequently and may involve more than one dimension at a time. Sustained G-loading can currently only be simulated in centrifuge devices specially designed for this purpose.

Unlike translational motion, rotational motions generally occur with more frequency in flight since they are the normal result of primary control inputs in aircraft pitch (*y*-axis rotation), roll (*x*-axis rotation) and yaw (*z*-axis rotation). These whole body motion sensations are typically experienced *after* those provided by the external visual scene or by the aircraft instruments. This is because visual motion sensory thresholds are much lower than whole body motion sensations so the pilot will generally always detect visual motion stimuli first (provided the visual motion cue is available and within the pilot's visual field).

While either translational or rotational forces can be imparted along each of the aircraft axes individually, the operation of an aircraft will more often induce motion accelerations or decelerations in more than one axis at a time.   For example, an aircraft in a level turn usually results in the pilot experiencing translational motion in the *z*-axis and rotational motion in the *x*-axis within the same maneuver.   Similarly, increases in aircraft pitch angle results in some rotational motion along the lateral or *y*-axis as well as some translational motion along the vertical or *z*-axis. Complicating the issue further is the fact that the pilot is usually seated at a distance from one or more axes of the aircraft. As a result, motion force effects on the aircraft axes will not necessarily be directly translated to the pilot.

Maneuvering and Disturbance Motion

How the pilot actually experiences whole body motion cues has given rise to a distinction between two types or classes of motion, maneuvering and disturbance (Guedry, 1976). *Maneuvering* motions are those motion forces that are the result of pilot actions in controlling the aircraft. These actions may be the result of pilot inputs through the primary control systems such as a yoke or stick control or the result of control inputs from rudder pedals. They may also occur as a result of secondary control inputs such as power, brakes, flaps or landing gear operation. Motion cues in the maneuvering category are always the result of pilot control inputs.

Disturbance motion cues are those cues which are the result of factors other than pilot control behavior. Typically, such disturbances are the result of aircraft system failures or environmental conditions such as weather. Examples of the former include engine, flap, and gear extension failures and examples of the latter include turbulence and wind shear encounters. Such motion cues tend to be unexpected and may serve as alerting stimuli for conditions requiring immediate attention. Disturbance motion cues are more likely to be brief in duration when compared to maneuvering cues and may be much more intense. For example, disturbance motion from turbulence can impart very high, albeit brief, G-loading on the aircraft that would not occur in normal, non-aerobatic flight operations. In some cases, disturbance cues may be the first indications of potentially unsafe conditions.

Sensing Motion

A pilot senses whole body motion forces through three sensory channels: the vestibular system, the cutaneous or tactile sensory system and the proprioceptive system. The vestibular system provides the pilot with the sensations of movement from translational and rotational *accelerations* of the aircraft. The vestibular system is composed of two components, the semicircular canals and the otolith. The semicircular canals respond to angular accelerations in all three body axes. The otolith, composed of the utricle and saccule, responds primarily to linear accelerations imposed on the body and the system and is most responsible for postural cues including tilt cues. The vestibular system, located in the inner ear, has a threshold of acceleration or change in velocity which must be exceeded in order for a given motion to be sensed by the pilot. For translational (or linear) accelerations, a typical threshold for pilot perception is 10 cm/sec² for oscillating linear motion (Howard, 1986). This is approximately .01 G of acceleration or deceleration. The sensory thresholds of rotational or angular accelerations vary depending on how the sensation is measured. Simple reporting of the onset of motion by a pilot (first report) will range between 0.44 and 0.80 deg/sec². Other measures involving the onset of illusory visual motion, such as the oculogyral illusion, yield a much lower threshold of between 0.10 deg/sec² to 0.12 deg/sec² (Clark and Stewart, 1968). The first report threshold values are probably more useful for the simulator platform motion designer since they directly reflect the pilot's subjective motion experience in the absence of any visual stimuli. The latter measure, which uses the oculogyral illusion, reflects the strong influence visual stimuli have on the perception of whole body motion.

A notable characteristic of the vestibular system is its prolonged after effects. The semicircular canal system contains a fluid mass which has built-in inertia. This mass makes the system somewhat slow to respond to the onset of acceleration forces and somewhat slow to stop responding once the acceleration has ceased. These after effects are known to contribute to pilot *spatial disorientation*. This is the inability to correctly assess the attitude of the aircraft, which results from the pilot's reliance on these same whole body motion stimuli. The avoidance of spatial disorientation is one of the reasons that pilots are trained to ignore whole motion cues in instrument flight and to attend solely to their aircraft instruments for attitude control.

While fairly significant accelerations are needed to stimulate the vestibular system, pilot motion sensation is not dependent solely on this system. Motion can also be sensed as pressure or vibration on the surface of the skin, typically from the seat pan, backrest or other components of the pilot's seat. These sensations undoubtedly gave rise to the 'seat-of-the-pants' expression that early pilots gave to their strategy of controlling an aircraft. The pilot's buttocks and lower back that contact the seat are sensitive to pressure and vibration at certain frequencies and amplitudes. Sensation of these stimuli will depend greatly on the area of the pilot's anatomy that is stimulated. For example, the threshold for sensing vibration from a 200 Hz sinusoidal signal in the buttocks area is achieved at 30 micron amplitude, while the lower back requires only 4 microns. These values suggest that the cockpit seat can transmit some useful disturbance and possibly some maneuvering cues to the pilot.

In addition to vestibular and cutaneous sensations, pilots also experience motion cues from proprioception. Proprioception is the result of feedback from sensors in the pilot's joints and muscles, which provide information regarding the position of the limbs relative to the body. For example, rotation of the aircraft about the longitudinal axis places forces on the pilot's arms which move them outward from the body. More dramatic sensations occur during negative g-loading when the limbs 'float' away from the body. Generally, proprioceptive cues play a significant role in motion perception only at higher levels of translational or rotational aircraft motions, as might occur in aerobatic or air combat maneuvering.

As discussed in an earlier chapter, all of the above nonvisual motion cues may be supplemented or even supplanted by strong motion cues coming from the visual scene outside the cockpit window. In the absence of such visual cues to motion, however, whole body motion cues will serve as the primary means by which a pilot senses aircraft motion. Therefore, whole body motion cues will tend to play a more significant role in flight simulation when no visual scene display is provided.

Motion Simulation Technology

Whole body motion simulation is inherently difficult to create on the ground since it requires the acceleration of the flight simulator (and the pilot). This acceleration requires that the simulator be moved over a distance—the greater the acceleration, the greater the displacement of the device that is required. With the exception of centrifuges, powered sleds and some research devices, significant linear accelerations on the order of 1.5 Gs or more, are not simulated in ground-based devices. For pilots who are likely to experience high G-loading as a part of their operating environment, only specialized devices or the aircraft itself are capable of delivering the kinds of G-loads that the pilot might encounter.

Angular accelerations, which are produced by rotation about the aircraft's three axes, only require movement of a device around a given point in space. For this reason, angular accelerations are less difficult to simulate than are linear accelerations. It is theoretically possible to provide all of the angular accelerations that could be experienced in advanced aerobatic maneuvers without displacing the simulator device from a fixed position on the ground. However, with the possible exception of rotations about the aircraft's longitudinal axis, most aircraft motions involve combinations of rotational and translational motions. For example, turning an aircraft involves both angular acceleration about the x-axis, some amount of yaw acceleration as well as some Gz-loading. The complex motion cues which accompany pilot control inputs generally require that a flight simulator be capable of providing some amount of both linear and angular acceleration cues.

The requirement to provide both types of motion cues inevitably involves design compromises. The accelerations provided are necessarily limited due to the physical constraints imposed by the size of the simulator facility. Apart from centrifuge systems, motion cueing systems cannot provide sustained linear accelerations (G-loads). With few exceptions, training and research simulators can provide only small and relatively brief accelerations. The high cost of very large and complex motion cueing systems, which are capable of providing much more realistic motion cues, limits their use to a few research facilities.

Both training and research simulator motion cueing systems can be best described as the end products of a series of necessary compromises. The compromises result in system designs that provide only a fraction of the motion cues that pilots might experience in actual operations. Compared with visual scene and sound simulation components, motion cueing systems are unlikely to ever be able to completely reproduce the motion accelerations possible in the operational environment. Because of its widespread use in more advanced training and research devices, motion platform technology will be described in more detail in the following sections.

Motion Platforms

Motion cueing systems are most commonly designed to move the entire simulator in one or more axes. The term *platform motion* describes systems that are intended to move the flight simulator cab in its entirety. The means by which this is accomplished varies widely, but the motion drive systems are typically composed of gantries, powered tracks, and hydraulically or electronically-driven rams. The Vertical Motion Simulator (VMS) located at NASA–Ames Research Center combines several of these technologies in the largest vertical motion platform system in existence (see Figures 3.2 and 3.3).

In Figure 3.2, the simulator cab and platform motion components are shown in detail. Pitch and roll actuators are revealed at the base of the simulator cab. Note that the vertical platform contains tracks which are used to provide linear accelerations in the simulator cab x or y axes. In Figure 3.3, the simulator cab is shown mounted on the vertical platform inside the VMS shaft. The platform moves up and down tracks imbedded in the shaft walls. As its name implies, the VMS was designed specifically to provide the largest amount of vertical motion travel possible within an enclosed facility. This facility is ten stories high, allowing nearly 22 m of vertical platform travel. The height allows the VMS to provide vertical accelerations of nearly 6.8 m/sec² or the equivalent of 1.75 Gz.

As impressive as the VMS is in providing Gz loading, it can provide only a fraction of the maximum design loads of typical civilian (4 Gz) or military (9 Gz) aircraft. However, the VMS is capable of providing the much lower Gz loading of normal aircraft maneuvers, including those likely to be experienced by large transport aircraft in the critical approach and landing maneuver (Bray, 1973). This makes the VMS a particularly useful research device for investigating large transport handling characteristics in the takeoff and landing phases of flight.

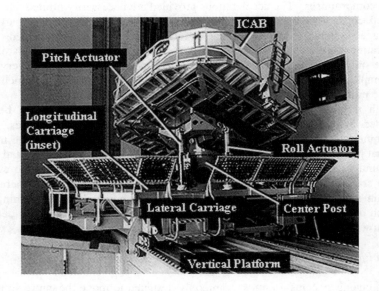

Figure 3.2 NASA VMS Simulator Cab and 6-DOF Platform
(Image Courtesy of NASA)

Figure 3.3 NASA VMS
(Image Courtesy of NASA)

While large motion platform systems such as that of the VMS can serve specific research needs, the cost of such systems are prohibitive for pilot training and evaluation. The development of a platform motion system that could provide at least some motion cueing in all aircraft-flight axes at a more reasonable cost was pursued relatively early in the history of flight simulation. Such a system had to accommodate the limited physical facilities available in most pilot training environments as well as the limited budgets of training organizations. It would also have to be extremely rugged and easy to maintain if it were to tolerate demanding crew training schedules.

The solution to the problem of an economical motion platform system came in the form of what is now termed the 'Stewart platform'. The Stewart platform motion design architecture is by far the most prevalent of all motion systems used for pilot training and is found on many research simulators as well. The platform is typically configured as a 6-DOF synergistic system that uses a set of six hydraulically or electronically-driven rams or jacks. The jacks are connected to the base of the simulator as shown in Figure 3.4.

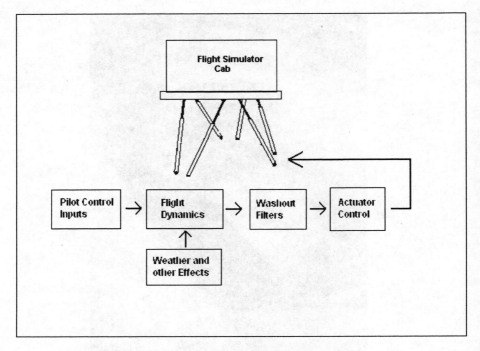

Figure 3.4 Illustration of the Stewart Motion Platform

The platform attachments of the jacks allow the jacks to move laterally by means of spherical joints. Owing to the limited displacement of the platform jacks (approximately 1 to 2 m), the physical interdependency of these jacks and the complex motions of the simulated aircraft, substantial limits are placed on the motion cues that can be produced. The restrictions resulting from the interdependency of the jacks alone is estimated to be approximately 50 to 60 percent of the displacement possible if the individual jack sets operated independently (Rolfe and Staples, 1986). The severe restrictions on platform motion cueing that result from the synergy of the Stewart platform actuators can have dramatic effects on the motion cues actually produced. Additionally, washout filters, which are needed to keep the commanded actuator movements within excursion design limits, can also introduce motion artifacts. (An example of the Stewart platform installed on a B-747 simulator is shown in Figure 3.5.)

Figure 3.5 B-747 Flight Simulator
(Image Courtesy of NASA)

Limitations of the Stewart Motion Platform

The significant limitations of the Stewart platform motion architecture are dramatically illustrated by a study of the platform system's response to simulated aircraft motion resulting from engine failure on takeoff in a Boeing 747 (Nahon, Ricard and Gosselin, 1997).

Engine failure on takeoff, particularly during the first segment climb, is a serious problem for any aircraft but is especially so for large transport aircraft with wing-mounted engines. Not only does the loss of an engine mean loss of power in a flight-critical situation, but for these aircraft it usually means at least temporary loss of some directional control and the potential for an aerodynamic stall of the effected wing – often fatal at low altitude.

Recognition and recovery from engine failure on takeoff is an important task for the pilot and is an integral part of all pilot simulator training syllabi. For the simulator motion platform designer, outboard engine failure in takeoff and first segment climb on a large transport aircraft like the B-747 is a significant challenge. This is partly due to the platform excursions required as well as the high pitch angle (around 20 deg) typical of large aircraft during takeoff and the initial climb segment. The high pitch angle limits the amount of simulator motion that can be applied in other axes.

The performance of the Stewart platform in this critical training task is therefore of considerable importance. Using the data of Nahon, et al. (1997), the performance of the Stewart platform was assessed by calculating the specific linear and angular motions provided by the platform as a percentage of the motion forces that would occur in the actual aircraft operations under similar circumstances.

The Nahon, et al. (1997) motion analysis is for a right outboard engine failure about 6 sec after rotation.   A comparison between the aircraft and the motion platform's peak motion forces at and shortly after engine failure reveal the following: the Stewart platform with the classic washout filtering system provided about 55 percent of Gx forces of the aircraft, 54 percent of the Gy and 29 percent of the Gz loads. The platform provided only about 15 percent of the roll rate of the actual aircraft, 19 percent of the yaw rate and 50 percent of the pitch rate. Additionally, the platform frequently provided washout motion in the *opposite* direction of the aircraft in four of the six degrees of freedom.[8] Only longitudinal linear motions and yaw angular motions were consistently provided by the platform in the same direction as the aircraft motion.   This peculiar platform motion behavior is a consequence of washout filter commands required to maintain the platform within its excursion design limits.

Since the motions in this example are a consequence of an aircraft malfunction (engine failure) and not pilot control, they are considered disturbance motions. Pilot control inputs to maintain safe aircraft speed and directional control would typically follow immediately upon the pilot's correct recognition of the engine failure condition.   If it is assumed that the motion cues associated with engine failure are the primary alerting cues that the pilot must learn in training, then there is a legitimate question as to whether the limited, low fidelity disturbance motion cues provided by the Stewart platform are sufficient to provide the pilot with adequate levels of motion cueing.

Even more important is the potential for the introduction of motion cueing artifacts that may be introduced by the platform washout filtering system. The washout motion, necessary to operate the platform within its limits, must be provided at levels below the pilot's sensory threshold for the linear or angular accelerations that might be produced by the washout motion. Analyses of Nahon, et al. (1997) washout motion data suggest that suprathreshold accelerations may indeed occur with the washout system and that the creation of motion cueing artifacts with the classic motion washout filtering used in the Stewart platform is a distinct possibility.

The Nahon, et al. (1997) data illustrate the inherent problems in synergistic motion cueing systems such as the Stewart platform. In order to develop a platform system that could provide motion cueing in a highly constrained workspace and within the cost constraints of the pilot training market, the motion platform cues provided by these systems are necessarily very limited. Moreover, the motion cues provided are not necessarily consistent across all flight conditions.   Due to the

---

[8] It is a common design strategy to allow for a repositioning of the platform using what are hoped to be platform movements that are too low to be sensed by the pilot.

synergistic nature of these platforms, the motion cues produced by an engine failure in level flight in the simulator could be significantly greater than those provided by an engine failure in the first segment climb, due solely to the constraints of the motion platform itself.

Adding to the problem of motion platform performance is the wide variation in motion cues that different aircraft impose on pilots under different operational environments. Despite these differences, the same constrained synergistic platform architecture is used for pilot training of all aircraft types. For example, the motion cues are quite different for engine loss in a single-engine aircraft compared with a multi-engine aircraft. They also differ for aircraft of the same weight category which have centerline or fuselage-mounted engines when compared with aircraft with wing-mounted engines. Many more such examples could be given for other aircraft and their operating conditions. The point to be made is that the synergistic motion platform, which has become a standard for all civilian FFS and for most military aircraft simulators, will differ widely in its ability to produce useful motion cues to the pilot across a variety of aircraft types and operating conditions. The one-size-fits-all design strategy of the Stewart platform is unlikely to produce the level of motion fidelity that justifies its role as the *de facto* standard.

The above analyses are an attempt to illustrate the problem created by placing cost and other factors above the needs of simulation effectiveness. The Stewart motion platform is a compromise design that has become a standard for motion platform systems. The use of a standard design is highly desirable in simulation component technology provided the technology is effective. In the case of motion cueing, the standard design may not be the most appropriate or even the most cost-effective. Only a detailed analysis of the full range of aircraft operating conditions and the motion cues they produce, coupled with an analysis of alternative motion platform architecture capabilities, can provide the necessary information on what kind of motion cues a particular platform design should provide for a given aircraft type or category.

The Role of Motion Cues and Simulator Motion Fidelity

It is important to note that whole body motion cues are not the only cues available to the pilot. For maneuvering motion cues in visual flight conditions, the pilot is likely to rely most heavily on his or her visual senses for vital feedback in controlling the aircraft. Not only is the visual scene a richer source of information for aircraft control, but the pilot's visual processing system is much more efficient than any of the non-visual processing systems available to the pilot. In instrument conditions, where no out-the-window visual information is present, pilots must depend on information from their instruments for maneuvering control feedback. They must learn to *ignore* whole body motion cues under these conditions in order to avoid spatial disorientation. While the training effectiveness of motion cueing systems will be discussed in a later chapter, the analyses of such systems' potential role in aircraft flight control described above suggest that the maneuvering motion cues provided by motion platform systems are of questionable training value.

Disturbance motion cues are, however, another matter entirely. Since they are not a product of pilot control inputs, they do not serve as a potential control feedback cue. While maneuvering motion cues may be an expected result of pilot control inputs, disturbance motion cues are not. Disturbance motion, therefore, plays a different role in piloting tasks than maneuvering motion.

The most important role that disturbance may have, and for which its simulation may be readily justified, is in its alerting function. As a primary cue for aircraft system failure, disturbance motion cues may be critical to efficient recovery from system failures. In the case of engine failure, for example, disturbance cues may prove extremely valuable where the disturbance motion is the *primary* cue to engine failure. In other situations, disturbance cues in the form of airframe buffeting and vibrations may be the primary cues to the failure of flap or landing gear operation. These and other instances of disturbance motion cues where the motion cue is the primary source of aircraft state information are essential for realistic flight simulation as well as being a necessary component of pilot training.

Disturbance motion cues also serve other functions and these are generally classified as *environmental* rather than as aircraft state cues. Environmental disturbance motion cues are those resulting from, for example, weather phenomena including turbulence and wind shear. They can also be produced by encounters with non-weather phenomena such as bird strikes. In ground-operations, they can be produced by runway contamination, runway and taxiway imperfections or anomalies, by the aircraft wheels leaving the runway or taxiway surface and by encounters with objects or debris on the runway or taxiway. Once again, where the disturbance motion cue is the primary cue, its accurate simulation is not only warranted, but imperative. Environmental disturbance motion cues may provide essential information that the trainee pilot needs to learn to recognize and that the experienced pilot needs for adequate situation awareness.

Vibration

A particular form of whole body motion can serve in either maneuvering or disturbance motion cue categories. *Vibration* is a very high frequency form of motion stimuli that is transmitted through controls, cockpit structures, and the surrounding airframe. Because the pilot senses these vibrations through structures and materials, the original vibration stimuli can be greatly attenuated or magnified before reaching the pilot's senses.

Vibrations imposed on the pilot have a variety of sources. The most common sources of low frequency vibrations are engine operations, flap and landing buffeting, runway and taxiway imperfections, and turbulence. Changes in vibration frequency or amplitude, which are a product of pilot control inputs, are by definition maneuvering cues. The pilot may come to expect changes in vibration or what are termed *vibrotactile* cues as a normal result of flight operations and respond accordingly. Indeed, normal vibrations may become so routine as to become habituated – aircraft vibrations become a form of background noise in flight. When habituated, vibrations may become only attended to when they are absent or when the vibration intensity or frequency changes unexpectedly. Under these conditions, vibrations can serve the same function as the disturbance motion cues discussed previously. That is, they may serve as important alerting cues to changes in aircraft or environmental states.

Humans are most sensitive to vertical vibrations in the 5 to 16 Hz range and to horizontal vibrations between 1 and 2 Hz (Sanders and McCormick, 1993). Vibration frequencies between 10 and 25 Hz will bring on discomfort as well as fatigue. Most aircraft vibrations are less than 30 Hz with amplitudes on the order of 0.05 rms G for large transport aircraft (Stephens, 1979).

Simulating Vibrations

Low amplitude, high frequency vibrations are well within the capabilities of most motion platforms including the Stewart platform. Indeed, only minimal displacement of a motion platform is necessary to deliver vibrations of this nature. In a study of motion platform effectiveness for a B-727 aircraft, for example, only 0.64 cm of platform displacement was needed to simulate operationally realistic vibrations (Lee and Bussolari, 1989).

The nature of these vibrations suggest that not only can they be provided by very limited platform motion devices, but that alternative devices for their simulation might be just as effective. As vibrations are primarily sensed by the pilot through the pilot seat, low frequency and low amplitude vibrations can be applied to the seat by means of seat shaker systems. An electric motor connected to a cam system controlled by the simulator host computer can provide vibration cues at a very low cost. However, such devices are only appropriate for simulation of vibrations at lower amplitudes (e.g., 0.01 G). Higher amplitude excursions of the pilot seat independently of the rest of the simulator cockpit would adversely affect the design eyepoint of any visual scene simulationprovided, the ability to read cockpit instruments and the ability to operate cockpit controls. The effects of seat vibrations are also limited to vibrotactile stimulation. They would have minimal or no effect on the pilot's vestibular system.

Alternatives to the seat shaker include the use of large subwoofer speakers which transmit very low frequency vibrations through the seat pan. While anecdotal evidence suggests that these systems may provide some of the range of vibrations needed in flight simulation, their actual effectiveness is not known. The effects of high amplitude, low frequency sound transmissions on other components of the simulation are also unknown. Factors such as seat material and construction will significantly affect the utility of such devices in producing the necessary amplitude and intensity of vibrations.

Devices for producing these vibrotactile cues which do not employ full platform motion are not only much less costly, but also reduce the maintenance and extend the useful life of the simulator. This is due to the fact that prolonged exposure to vibrations is deleterious to sensitive electronic components of a simulator. Isolating the vibrations to the seat avoids exposing these sensitive components to the potential damaging effects of vibration.

Vibrations, as with other motion cues, may provide valuable information to the pilot regarding the state of the aircraft of the environment within which the aircraft operates. Some of these are provided in modern simulators, but only in those equipped with platform motion systems. Simulation of these cues by other, less expensive means needs to be more rigorously investigated for use in training and research devices.

Summary

Whole body motion affects the aircraft pilot through several sensory systems including vestibular, tactile, and proprioceptive. The primary means of providing whole body motion stimuli in flight simulators is by means of motion platforms, specifically the Stewart platform. It is suggested that the maneuvering motion cues provided by these platforms probably are of little or no value to the pilot. However, disturbance motion cues are of potentially great value in flight simulator motion simulation, particularly if they serve as primary cues to unexpected changes in aircraft or environmental states. Alternatives to platform motion systems are also discussed and it is suggested that their efficiency in providing motion cues needs to be investigated further.

Chapter 4

Handling Qualities and Control Loading

Aircraft Control

The control of an aircraft can be divided into three states: manual, semi-automated, and automated. In manual control, aircraft state parameters such as airspeed, altitude, rate of climb or descent are under the direct manual control of the pilot through the manipulation of *primary* and *secondary* controls. Primary controls include the use of yoke (or control stick) to change aircraft pitch and roll, rudders to control aircraft yaw, and throttles to control aircraft thrust. Secondary controls include the use of flaps and related devices to control wing lift properties. They also include elevator and aileron trim controls and braking systems for ground operation.

Pilot manual control can be further divided into closed loop and open loop. Closed loop control occurs when a pilot must maintain continuous input control in response to continuous feedback. Closed loop control occurs in most primary control situations where the pilot is, for example, adjusting aircraft pitch attitude and thrust to maintain a specific rate of climb or attempting to hold a constant heading. The control inputs are adjusted as needed in response to information from the visual scene or aircraft instruments to the pilot's control inputs. The continual control adjustments in response to changes in information form a closed loop. In open loop control, a pilot can set a control to a prescribed value precluding the necessity of continuous feedback and control input monitoring. Examples of open loop controls include the setting of takeoff trim and takeoff power values. Much of the complexity in simulating the handling characteristics of aircraft are devoted to closed loop control systems where the need for accurate simulation of control forces is imperative.

Automated Flight Control

Regardless of whether closed loop or open loop control is employed, both categories of pilot control have the ultimate goal of achieving certain target aircraft attitude, airspeed, rate of climb or descent, or altitude. With the introduction of modern computer technology, automation of most or even all of the pilot's manual control tasks is now a reality. Most aircraft, including small general aviation aircraft, now have at least some capability to provide automated control of aircraft state parameters. *Semi-automated* control systems are usually limited to maintaining airspeed, altitude and heading or directional control by providing the primary control inputs that are normally a part of the pilot's duties. These 'autopilots' remove much of the manual workload required for primary control of an aircraft, but many secondary control tasks remain. Moreover, such systems do not take over the more complex flight management tasks such as navigation nor do they have any capability to land the aircraft without pilot intervention.

Fully automated systems have now been introduced in many aircraft, both civil and military. Such systems have now largely replaced the need for pilot primary or secondary control inputs to control the aircraft. Additionally, through the use of advanced Flight Management Systems (FMS), pilot navigation tasks have also been reduced or eliminated. With minor exceptions, these automated systems are capable of completing all phases of flight without any pilot intervention whatsoever. Such systems have the potential for significant increases in the efficiency of aircraft operations. Unfortunately and unavoidably, because these automated systems are replacing the need for pilot manual control, pilot manual control skills will tend to degrade over time if no recurrent training is provided to maintain these skills. The flight simulator will play an increasingly important role in maintaining pilot manual control skills. Regularly scheduled refresher training for selected manual pilot skills is already a part of commercial airline flight simulator training for aircrews. This training is particularly important to maintain flight-critical skills such as those involved in takeoff and landing. Increasingly, simulators will need to provide high fidelity handling characteristics in order to maintain pilot manual control skills.

Simulating Flight Control

Clearly, the application of manual control skills in 'flying' a simulator requires that the simulator respond to pilot control inputs in a manner closely comparable to the aircraft being simulated. Figure 4.1 illustrates how a flight simulator would attempt to re-create the control feedback loop necessary for the simulation of primary flight controls such as the yoke and rudders. When the pilot initiates a control input in a real aircraft, the aircraft's control structures provide a counterforce as the control mechanics carry out that input. In small aircraft, these control inputs act on a collection of cables, pulleys and levers to move control surfaces on the wing and tail sections. When the aircraft is parked or moving at very slow speeds where there are no aerodynamic forces acting on these control surfaces, the loading on pilot controls are said to be *static*. Simulating these static control loads does not require aerodynamic or other complex modeling since the loads on the control surfaces do not change dynamically. In a flight simulator, this type of control loading can often be simulated by relatively simple, passive systems including heavy duty springs, counter weights or friction devices that are connected to the control device. These loading systems can provide some of the 'feel' to the pilot approximating the control forces that would be expected in the real aircraft.

As seen in Figure 4.1, a simple control loading system is designed to provide counter forces to pilot control inputs. In this case, yoke control inputs in both roll and pitch inputs correspond, respectively, to forces on the aircraft's aileron and elevator control surfaces. In a two-axes control system, such as would exist in this example, two sets of control loading systems are provided. The two different sets are needed because the forces required to simulate movement of the respective control surfaces (aileron and elevator) have different characteristics. When operated in the simulator, the displacement of the controls are detected by two separate position sensors, one for the yoke aileron inputs and one for the yoke elevator inputs The position sensors are needed because the effect of control displacement differs depending upon the airspeed of the aircraft. As speed increases, the control surfaces in the aircraft become more sensitive to control inputs so that the same amount of control displacement will have a much greater effect on the aircraft at higher speeds than at lower speeds.

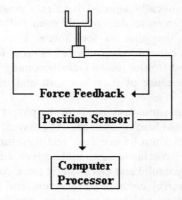

Figure 4.1 Simple Control Loading System

Thus, even with highly simplified control loading, the aircraft simulated will still respond correctly with respect to the amount of *displacement* of the control. This is provided by a computer processor which amplifies or deamplifies the control input signal in accordance with the simulator's flight control model.

In simple control loading systems, the amount of force required by the pilot to effect the same amount of control displacement remains the same regardless of aircraft state or any other conditions. This is because the fundamental characteristic of the simple control loading device remains the same regardless of simulated flight conditions.

Many flight training devices use some form of simplified control loading device. The most common such devices use heavy springs to provide the necessary forces that will resist pilot control inputs. As the amount of displacement of the control increases, the forces required for additional displacement also increase as the spring reaches full extension. Spring-based systems, sometimes called 'spring-centering' systems, are a low-cost answer to control loading. Regardless of the system used, they should be able to provide forces within ten percent of the reference control force values of the aircraft in order to avoid pilot detection of force loading differences between the simulator and the aircraft (Biggs and Srinivasan, 2002). Large differences between simulator and aircraft control loading can lead to problems when the pilot attempts to transfer simulator control skills to the aircraft.

The simplified control loading scheme has been improved in recent years with the advent of force feedback systems using processor-controlled, electronic motors. The motors provide force feedback by means of gears or belts which apply torque forces to pilot controls, such as the yoke control illustrated above. In the above example, two separate motors would be provided, one for each control axis, both under microprocessor control. The advantage of this type of force feedback system is that it allows for *active* control of the feedback system. Not only can the levels of force feedback be dynamically adjusted to reflect changing flight conditions, but the systems can apply forces to the controls even without pilot inputs. These command forces allow the pilot to feel forces from turbulence or runway anomalies through the controls themselves. Force feedback systems using microprocessor controlled, torque motors are becoming increasingly popular in flight training devices because of their low cost and improved control loading flexibility.

More advanced flight simulators used for training and research incorporate much more complex control loading systems. These were developed in an attempt to provide more realistic control force cues and therefore more realistic aircraft handling characteristics. To achieve these goals, advanced control loading systems include more complex hydraulic and electrostatic force control systems as well as more complex and powerful computer processors and software modeling. An example of one of these control loading systems, the hydraulic control loading system for NASA's Vertical Motion Simulator, is shown in Figure 4.2. The figure illustrates a number of features that are absent from simpler control loading systems. One such feature is the ability to detect and measure the amount of forces actually applied by the pilot to the control system as well as the velocity of control displacement in addition to the absolute displacement of the control.

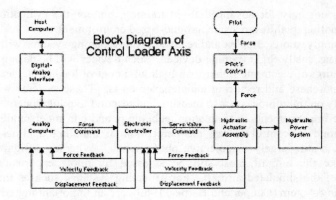

Figure 4.2 Hydraulic Control Loading System
(Image Courtesy of NASA)

The complexity of hydraulic control systems is matched by their ability to produce considerable forces on pilot control systems. This is illustrated in Figure 4.3 which shows the rudder pedals used in the NASA VMS. Note the heavy hydraulic pistons attached to the base of the rudder pedals. The forces of these pistons could easily overwhelm pilot control inputs if that were desired.

Figure 4.3 Hydraulically Controlled Rudder Pedals
(Image Courtesy of NASA)

The development of hydraulically controlled loading systems ushered in the era of high fidelity, dynamic control loading in flight simulators. Dynamic control loading meant that control force simulation could be changed to realistically simulate changes in operational flight conditions. These include changes in control forces due to system failures such as hydraulic leaks, changes in control feel due to wing and control surface icing, and other events that might affect control feel. The introduction of these active systems also allowed for the simulation of highly realistic aircraft handling characteristics. This, in turn, allowed greater use of flight simulators, not only for advanced flight training, but for the evaluation of new aircraft handling qualities in a safe, ground-based environment.

Increasingly, more complex and realistic control loading systems will find their way in to less costly flight training devices. Such devices will be designed around some variation of electronic rather than hydraulic control loading devices in order to reduce purchase and recurring maintenance costs. These systems will depend increasingly on microprocessors to measure and respond to pilot inputs.

Despite improvements in simulation technology and our understanding of how pilots respond to control forces, the perception of handling qualities in flight simulators is still largely a subjective phenomenon. Whether a flight simulator handles like the aircraft must still be assessed by pilots experienced in the operation of the simulated aircraft. Formal rating systems, such as the Cooper-Harper rating system (Cooper and Harper, 1969), are an important ingredient in the assessment of simulator handling qualities and should be used as the means of assessing the perceived fidelity of simulator handling characteristics.

Proprioceptive and Tactile Cues

Simulating flight control is not limited to recreating control forces, however. Although of importance to eliciting the proper manual control responses from the pilot, the gross perceptual-motor control feedback normally associated with yoke or rudder inputs are only part of the control simulation required. Both proprioceptive (kinesthetic) and haptic (touch) sensory feedback occurs as a part of pilot control behavior. The proprioceptive feedback occurs as a result of changes in the position of limbs and joints during the operation of control systems. These include primary and secondary controls as well as a variety of other instrument switching and control operations.

Most proprioceptive cues are those associated with control operations where visual or other sensory feedback is absent. For example, during visual flight operations, most power control operations are commonly executed with no visual reference to the position of the control. Movement of a control, such as a throttle lever or flap handle, often depends on proprioceptive feedback. For this reason, design and placement of these controls in the simulator should correspond closely to the aircraft control configuration being simulated. While the accuracy of proprioception varies depending on the direction of motion and distance from the body centerline, differences in control position on the order of 1 cm can be readily discerned by proprioception alone (Brown, Knauft, and Rosenbaum, 1947).

In addition to proprioceptive cues, pilots will also use haptic or touch sensations to discern control shape and other defining characteristics. Haptic cues along with proprioceptive cues aid in discriminating controls when the pilot needs to be looking elsewhere. Typically, aircraft power, flaps, landing gear and other control levers have control knobs of distinctive shape, size, and even surface texture to aid in touch discrimination. Reproducing these specific control shapes and textures in the simulator is necessary to avoid the need for pilots to rely solely on visual discrimination among controls. Failure to provide these tactile cues for control discrimination adds unnecessary visual workload for the simulator pilot and may force the development skills inappropriate for the aircraft flight operations that the simulator is intended to represent.

For passive control simulation, where the simulator designer needs to reproduce the location, size and shape of a control, the problem is readily solved by using an aircraft control part or by fabricating a copy of that control. But for active control simulation such as the examples of control loading described earlier, technical challenges to realistic simulation can be considerable. This is particularly true for haptic simulation where the need is to create the sensation of touch, such as vibration or pressure. These sensations may occur routinely in the course of pilot interaction with controls or other cockpit components. Active haptic interfaces capable of simulating touch sensations have only recently been developed for use in medical training, such as for surgical procedures, and are now providing the sensitive touch feedback needed in operating telerobotic equipment. Lessons learned from these applications should help designers improve the realism of haptic sensations in flight simulators.

The sensation of touch is important for sensing pressure on the skin's surface. Such pressures come from a variety of motion forces including vibrations and G-loading. These pressures are most commonly felt through control surfaces and through the cockpit seat as these are the components of that most often come in contact with the pilot's body. Pressure sensitivity differs dramatically depending on which area of the pilot's body is affected. For example, sensitivity to pressures is significantly higher for the fingers and hands when compared to the lower back. At 250 Hz only 0.1 mm of indentation is needed for the sensation of vibration to be detected at the tip of the middle finger. This is far less than the indentation of 30 mm needed in the buttocks area (Sherrick and Chulewiak, 1986). As is typical with sensations to vibration stimuli, more pressure is required in order to elicit a response at lower frequencies. At 10 Hz of vibration to the middle finger, for example, pressures resulting in 10 mm of indentation are required in order to be detected. Additionally, much less surface area of stimulation is required to create the sensation of vibration at the fingertips than in the lower torso of the pilot. The single advantage of applying vibrations to the lower torso area in a simulator is that this area is in continual contact with the cockpit seat while the fingers and areas of the hand may only occasionally be in contact with controls, such as the yoke or control stick. This is especially true if the aircraft has an automated flight control system. In some flight regimes and particularly with autopilot use, a pilot will not be touching control devices or will do so very intermittently. Although far less sensitive than the hand or fingers, providing vibrotactile stimulation through the cockpit seat may prove to be more effective in simulating this aspect of aircraft flight.

Wherever the vibrotactile or pressure sensation is ultimately simulated, the decision does affect the type of simulation technology that can be employed. Motion platforms can, of course, provide high frequency vibrations to the entire simulator cab. While straightforward, the use of motion platform technology in this way is probably not cost effective if vibration stimuli are all that is desired of the platform. Control loaders, seat shakers and other similar devices are much more cost-effective if providing high frequency, low amplitude vibrations to a specific area is all that is desired.

Simulator Process and Control Delays

Flight simulators are complex, ground-based devices that attempt to create a synthetic, but realistic flight environment for pilots. This necessitates the development of complex electrical and electro-mechanical systems that process and control various components of the simulation. Inevitably, the systems take time to complete their process and control cycles and these delays may introduce artifacts into the simulation environment. In assessing the ability of a simulator to respond in the way in which the aircraft would respond, these *process and control delays* can significantly distort the perceived responsiveness of the simulated aircraft to pilot inputs. If the delays are too large, the pilot will develop control strategies that are inappropriate, even hazardous, when transferred to the operational aircraft. Note that simulator process and control delays are distinct from the *transport delays* that are incurred in aircraft control. Transport delays are represented, for example, by delays in instrument responses to pilot control inputs. Such delays are a normal part of the aircraft's operation. The process and control delays that are discussed here are artifacts that may be added to the normal aircraft transport delay by the simulator.

Process and control delays can occur in a variety of simulator components including visual scene simulation systems, motion platforms, control loading, or instrument displays. Most simulator process and control delays have occurred in the past because of slow computer processor capabilities. The older computer processors simply did not have the capacity to deal with large amounts of data in a short period of time. The resulting simulator process and control delays created a variety of problems not only for simulator designers, but for those pilots who had to adapt to the simulator's peculiar behavior. Most of the more significant process and control delays have been eliminated from newer flight simulators because of the availability of very fast computer processors, but some transport delays still persist due to inherent limitations in electronic or electro-mechanical components.

Most of the concern today is on process and control delays that affect the update of visual scene displays. This is largely due to the very large computational demands placed on computing systems by very high resolution image generation and display systems. In general, visual display update rates under 50 msec are not considered a problem because delays at this level are generally beyond the level of human perception. Delays in the range of 50 to 80 msec are generally also not a problem except for those systems requiring very high levels of active pilot control, such as occurs in the manual flight control of helicopters (Wildzunas, Barron, and Wiley, 1996). Delays above 80 msec should be avoided if possible. Visual update delays of this extent are known to introduce handling quality artifacts into the simulation environment (Jennings, Craig, Reid, and Kruk, 2000; Levison, Lancraft, and Junker, 1979).

Summary

The simulation of aircraft handling characteristics has been a challenge for simulator designers from the beginning. Simple control loading systems which provide resistance forces to control input have subsequently evolved into very complex and sophisticated systems capable of continually monitoring and feeding back control loads to pilot inputs. Such systems can also provide other flight cues such as low amplitude, high frequency vibrations associated with a variety of aircraft and environment conditions. Newer, less costly force feedback systems using microprocessor and electronic torque motors will eventually bring high fidelity handling characteristics even to the most inexpensive flight training devices.

Chapter 5

Cognitive Fidelity and the Simulator Task Environment

Introduction

In the preceding chapters, the simulation of the visual scene, whole body motion, control loading, sound and related phenomena were discussed in some detail. Stimulating the pilot's sensory systems in a manner comparable to that which they would experience in the aircraft is perhaps the essence of what most would consider 'realistic' simulation. In analyzing the requirements for this sensory stimulation of the pilot, however, it becomes increasingly evident that much of what is provided in modern flight simulators is often unnecessary. Only selected elements of the visual scene that is available to the pilot at any time are actually needed to accomplish the flying task. While the aircraft is capable of highly complex motion, only a small segment of the motion cues available to the pilot are ever used. These examples and many others are illustrative of the active selection of information that occurs during the piloting task. It is the need to support the many and varied tasks that an individual pilot or an aircrew need to accomplish that is the real essence of flight simulator design. Beyond the provision of sensory stimuli, flight simulators increasingly need to provide a task environment which requires the pilot to exercise complex, *cognitive* skills.

Cognition and cognitive processing refers to a variety of higher mental functions such as memory, attention, and symbolic reasoning. These individual cognitive processes often place great demands on the mental resources available to the pilot. Piloting skills, which are largely dependent upon these cognitive processes, are classified as cognitive skills. Cognitive processes represent a significant proportion of pilot skills, such as workload management, planning, communication, problem-solving and decision-making. Unlike some of the highly learned perceptual-motor skills a pilot may possess, multiple cognitive skills generally cannot be performed simultaneously. For example, a pilot will find it very difficult to simultaneously program a navigation computer, mentally calculate fuel remaining and negotiate a new route with air traffic control no matter how skilled the pilot may be at performing these skills individually. Regardless of the experience level of the pilot, these activities demand a great deal of the mental resources of a pilot and place heavy loads on the mental workload of the pilot. As aircraft and airspace systems become more complex and demanding, pilot cognitive skill development will play an increasingly important role in aviation safety.

In meeting the needs of pilot cognitive skill development, the flight simulation environment must provide not only high levels of perceived fidelity in such areas as visual scene simulation and handling qualities, but an environment conducive to the training and evaluation of pilot cognitive skills. In other words, not only a high degree of perceived fidelity, but also *cognitive fidelity* is required in flight simulators. Cognitive fidelity of the simulation environment derives from a combination of flight simulator design and the design of the task environment within which the pilot and the simulated aircraft must perform.

Cognitive Processes and Skills

A more detailed description of cognitive processes is needed in order to clarify their role in the design of the flight simulation environment. Cognitive processes are those higher mental processes which are engaged volitionally by the pilot and typically involve the control of attention and memory and the manipulation or processing of information in order to support performance of a particular task. One such cognitive function, *attention*, is a process of allocating mental resources, either voluntarily or otherwise. While a pilot cannot attend to more than one task at a time, a pilot can process information from more than one source at a time (Atkinson and Shiffrin, 1968) provided some of the processing can be done at a level that does not require conscious effort.

Developing the skill to efficiently allocate attention to various flying tasks is one of the key skills a pilot must develop. This 'multi-tasking' ability, as it is sometimes called, is a necessary skill in an environment where task demands are high and where task completion is constrained by time. A highly skilled pilot can normally perform routine mental tasks without this activity interfering with a well-learned perceptual-motor skill such as the coordinated turn. *Ab initio* pilots can be easily overwhelmed by the information processing demands of flying because many perceptual-motor skills still require conscious attention. When some of the more basic flying tasks, such as the coordinated turn, become learned to the point at which they become automatic, much less demand is placed on limited cognitive resources. Then other cognitive tasks, such as radio communications, become much easier to accomplish and much less likely to interfere with the learned perceptual-motor task. While skilled pilots develop the strategies needed to allocate attention to flying tasks that are most effective in terms of resource utilization, *ab initio* pilots require extensive training and guidance. Many of these are now described as resource management skills in the pilot training community. Mental workload measures are often used to assess the degree to which a pilot's available resources are being used in a given task environment (Hart and Staveland, 1988). Such measures can be used as indicators of the degree to which the flight simulator task environment is comparable in mental and physical workload to the actual flight environment.

Another cognitive process routinely used by a pilot is the storage and retrieval of information from memory. Human memory processes are usually subdivided into *working* memory and *long term* memory. Working memory as its name implies, is used as an intermediary point either during the process of transferring information to long-term memory or retrieving information from it. The most common example of working memory in action is the pilot's use of memory items for emergency checklists. Since some emergencies, such as engine failure, require immediate action, paper or electronic checklists cannot be used initially. Pilots instead use checklists committed to memory so that responses to the emergency are not delayed. The memory items are retrieved from long-term memory and placed into the pilot's working memory where they are used to guide pilot actions during the emergency. Because working memory has a very limited capacity, only 4 to 5 steps are typically committed to memory. Working memory is also subject to interference from information coming from other sources such as crew or ATC communications. These sources of interference can lead to memory errors of commission or omission. Generally, the more similar the information to that in the pilot's working memory, the greater the potential for interference.

While long-term memory has effectively unlimited storage capacity, it is also subject to errors. For example, if a pilot flies several different aircraft regularly, the memory items retrieved during the engine failure may not be appropriate for the particular aircraft being flown at the time of the emergency. As long-term memory retrieval is more a process of construction than of retrieving recorded events, a pilot may fill in a missing detail with information that seems logical at the time. This happens most often with tasks that are repeatedly routinely. A routing or clearance from ATC that has missing or changed information is 'filled in' with information that logically fits the conditions or has been given many times before. This 'construction' of reality may occur whenever there is a need by a pilot or aircrew to 'make sense' of prevailing conditions or information. Reliance on long-term memory of similar past instances of the event is a reasonably adaptive strategy, albeit potentially incorrect for the situation at hand.

Most flying tasks are organized into steps or procedures in order to minimize operational errors and accidents. Checklists and procedural manuals are developed to cover every conceivable or normal or abnormal event. But not every event or condition that a pilot encounters can be predicted. *Problem-solving* describes a set of cognitive tasks and cognitive skills which pilots need to apply to resolve anomalous situations. These include the ability to identify and prioritize problems, to test alternative hypothesis, and to apply risk management strategies to affect a desired outcome. Effective resource management and, where applicable, crew management, are part of the problem-solving process.

Decision-making is closely allied to problem-solving in terms of the cognitive processes used. Decision-making requires the pilot to make choices among a set of alternatives usually with incomplete information. Decision-making under conditions of uncertainty depends heavily on the pilot's knowledge of how prevailing or predicted conditions will affect the outcome. One of the more common decisions a pilot may have to make is the diversion to an alternate airport.

Diversions are often due to weather or aircraft equipment malfunctions, which make a landing attempt at the planned destination impractical or hazardous. Exercising this kind of operational judgment requires the pilot to collect and evaluate a variety of information from a variety of sources. Some information, such as weather, is less reliable and subject to change. Other information, such as fuel consumption rates may require time and effort to find or calculate.

For the simulation environment to provide a high degree of cognitive fidelity, the simulation designer and user must have a through understanding of the pilot's operational task environment. That task environment may differ widely depending on the type of aircraft operations for which the pilot is being trained or evaluated. As a consequence, the simulation of the same aircraft type may differ significantly from one operational environment to another.

Simulating the Task Environment

Unlike the sensory and perceptual-motor elements of flying skill that have dominated the design of flight simulators, flying tasks which depend on pilot cognitive skills and processes depend on creating a simulator task environment which requires their use. Flight simulator design has been historically driven by the need to reproduce the stimuli necessary for the physical operating characteristics of the aircraft environment. This includes the desire to reproduce at a level of realism, the visual scene a pilot views out of the cockpit windows, the control feel of the aircraft, the motions associated with aircraft flight, and other physical features of the cockpit, such as controls and instrument display. This emphasis on *physical* fidelity equates the quality of a flight simulator device with the capability of the device to present a physical replica of the aircraft cockpit environment; the more accurately the device physically replicates the real aircraft cockpit, the higher the fidelity.

However, the quality and effectiveness of a flight simulator is only indirectly related to its ability to physically replicate the real cockpit environment. It is the perception of that environment, the *perceived* fidelity of a simulator that really matters. What is important to simulation fidelity at this level is what the pilot needs to experience in the simulation. If the pilot *experiences* a control feel that is indistinguishable from the actual control feel of the aircraft, then the simulation will be successful. This is true regardless of how closely the control loading system actually replicates control feel as measured by other means.

Cognitive fidelity is simply an extension of the concept of perceived fidelity to higher mental processes. The measure of cognitive fidelity of a flight simulator is its ability to provide the task environment necessary for the pilot to exercise cognitive skills and processes comparable to those that will be experienced in the operational aircraft. For example, basic instrument flight training is a common task for which many simulators are well suited. In most cases, instrument trainers will provide much, if not all, of the perceived fidelity necessary to train and evaluate a pilot's instrument flying skills. They may not, however, provide the necessary task environment needed to support the training and evaluation of the cognitive skills associated with instrument flight. The radio communication studies conducted by Lee (2003) indicate that the levels of mental workload provided by flight simulators without realistic ATC communications simulation are significantly lower than the operational environment and may lead to inappropriate pilot behavior. Studies conducted on early use of flight simulators found that their task environments emphasized manual flight skills at the expense of crew and resource management skills (Ruffell-Smith, 1979). These studies led to the development of Line-Oriented Simulations (LOS) in an attempt to improve commercial airline pilot training by enhancing the task environment. Among these are the Line-Oriented Flight Training (LOFT) scenarios that are now commonly used in commercial airline training. These LOS environments are particularly well-suited to the development and evaluation of cognitive skills including workload management strategies, crew resource management as well as problem-solving and decision-making.

Cognitive Fidelity and Simulator Design

It is not necessary, of course, that all flight simulators be capable of creating a fully realistic LOS environment in order to provide a high degree of cognitive fidelity. What is necessary is that the device provides the essential components of the operational task environment that are needed to exercise a given cognitive skill or set of cognitive skills. This is best accomplished by applying cognitive task analysis (CTA) techniques to identify the cognitive processes involved in the particular flying task. Methods for conducting CTA are beyond the scope of this book, but are discussed elsewhere (see Schraagen, Chipman and Shalin, 2000). An example of how such an analysis might be accomplished and how it would impact simulator design is provided in the following sections.

Cognitive Task Analysis: Precision Instrument Approach

CTA, as with any task analysis, seeks to decompose a task into simpler elements in order to reveal cognitive functions and processes involved in the task. These analyses should be conducted for all relevant flying tasks, including crew resource management tasks if applicable, for the specific aircraft type or category, and for the full range of operating environments that are of interest. Crew composition, automation, weather and other operational factors affect the nature and complexity of flying tasks and the cognitive processes involved. For example, single-pilot, IFR operations in a small aircraft without any automation are among the most difficult of any in civil aviation, largely because of the high demands placed on the mental workload of the pilot.

Table 5.1 shows some of the major tasks required during the cruise, descent and approach phases for a single-pilot instrument approach. The most common cognitive activity throughout is that of allocating attention to the various tasks that need to be carried out. The pilot must be able to develop the skill necessary to efficiently allocate cognitive resources to various tasks in order to minimize errors across tasks. How this is accomplished for a given aircraft can vary considerably depending on the aircraft's handling characteristics, available avionics and automation, type of instrument approach, and weather conditions.

The second most common cognitive task in the example is the continuous monitoring of aircraft state parameters and of radio communications frequencies. Again, tasks like monitoring aircraft speed, altitude and other parameters will vary in difficulty depending on the level of automation and other factors mentioned previously. The monitoring involves a process of comparing the aircraft state parameters with those target parameters desired. The target parameters are typically stored in the pilot's working memory, which may be aided by reminder devices attached to the instrument (e.g. airspeed bugs) or otherwise displayed on the instrument panel. Monitoring of radio communications is also conducted continuously throughout the flight in order to assure reception of ATC directives and clearances. The ATC directives for the aircraft or ownship communications must be detected against a background of radio frequency chatter caused by communications to and from other air traffic on the same frequency. Ownship communications between the aircraft and ATC require only a small proportion of pilot cognitive resources when compared with other communications monitoring activities.

Certain planning and preparation activities are carried out, usually in the latter stages of the cruise phase and again in the approach phase of flight. These planning activities are conducted to review approach procedures, including approach minima and missed approach procedures. They involved a detailed analysis of approach, chart information, as well as available current and project weather information for the destination airport. Retrieval of procedural knowledge of how to conduct instrument approaches from long-term memory occurs at this stage, including information on proscribed and personal approach minima. Similar activities occur during the initial stages of the approach with reviews of procedures and approach minima.

Decision-making can occur throughout a flight, as well as before a flight. The most important of these decision-making events is the diversion decision. This decision-making process occurs most commonly during the approach to the destination airport, though it may occur much earlier in the flight. Along with all the other tasks, the diversion decision can impose a significant increase in the mental workload of the pilot. The decision involves both the decision to divert from the intended destination airport and a potentially additional decision of an alternate airport if more than one is available. Decision-making is among the most complex of cognitive tasks as it involves the collection and weighting of a variety of information from several sources. Information on weather, runway and airport facilities, fuel available and other information, needs to be analyzed in real-time and in the context of ongoing flight tasks. Weighing the importance, value and even reliability of information places a considerable strain on a pilot's working memory and on cognitive resources that may be needed elsewhere.

Table 5.1 Cognitive Tasks Performed During a Single-Pilot, Instrument Approach

	Task	Cognitive Process
Cruise	- Planning/preparation - Monitoring aircraft state parameters - Monitoring/processing radio communications	- Long term/working memory - Attention allocation - Working memory and attention allocation
Descent	- Monitoring/correcting aircraft state parameters - Monitoring/processing ATC radio communications - Monitoring of broadcast communications for terminal area information - Airport diversion decision	- Attention allocation - Working memory and attention allocation - Attention allocation long term & working memory - Decision making
Approach	- Monitoring/correcting aircraft state parameters - Monitoring of communication for ATC directives - Review approach procedures - Continue/discontinue approach - Airport diversion decision	- Attention allocation and working memory - Attention allocation - Long term/working memory - Decision making - Decision making

The analyses of cognitive functions revealed here is cursory as it is intended to only illustrate the issues involved in defining a simulator task environment, which would have a high degree of cognitive fidelity for this particular flight task. However, while the requirements for the *perceived* fidelity of, say, a visual scene are relatively straightforward, the requirements for the *cognitive* fidelity provided by a simulator's task environment would seem to be much less so. A reliable means by which one can determine the relative cognitive fidelity of a flight simulator needs to be developed. One means of doing so is to analyze the simulator's ability to deliver the needed information in a form and time comparable to that of the operational environment. This means that the simulator must increase the ability to create an information environment comparable to that of the aircraft's actual operating environment.

Simulating the Information Environment

Cognitive fidelity of a simulator's task environment means that the environment must support a pilot or aircrew's cognitive skills and processes in a manner and at a level comparable to that experienced in the operational environment. A corollary to this requirement is that the simulator task environment does not require cognitive activities that would not normally occur in the operational environment. Avoiding the development of simulator-specific compensatory skills applies to cognitive as well as to sensory and perceptual motor behaviors.

Cognitive activities can be considered as information processing activities in which the availability, timing, and format of the information have a direct and significant effect on how the cognitive activity is performed. Decision-making requires that information normally available to the pilot in the operational environment also be available in the simulated environment. For example, the diversion decision requires information on weather, fuel available and rate of consumption, airport facilities information, runway visual range minima, missed approach procedures, and other information. Part of the cognitive skills involved in the decision-making process is the ability use the various aircraft and ground-based resources efficiently and effectively in order to obtain the needed information in a timely manner. The simulated environment must be able to provide the opportunity for the exercise of these resource management activities by re-creating an information environment comparable to that of the operational environment. The nature of this information environment will depend on the particular aircraft being simulated as well as the type of operations for which the aircraft is being used. For example, electronic information systems dominate some aircraft environments while others still rely on information in paper form. The former systems markedly affect the speed and effectiveness of planning and decision-making activities. Ground-based resources also markedly affect how pilots develop and use cognitive skills. ATC clearances and terminal information provided on a piece of paper and handed to a pilot do not engage the same processes as those involved in processing information from radio communications. For a high degree of cognitive fidelity to be achieved in the simulated environment, information needs to be delivered in the same form as it is in the operational environment in order to assure that comparable cognitive activities by the pilot or aircrew will take place in the simulator.

The timing of information delivery as well as the rate at which information is updated also plays a role in the cognitive fidelity of a simulated environment. Since the operational environment is inherently dynamic, information about that environment is subject to change. Aircraft state parameters such as heading and airspeed are subject to continual change due simply to the disturbances impinging on the aircraft as it moves through the atmosphere. The timing of information and the rate at which it is updated can have profound effects on a pilot's cognitive activity. For example, for some aircraft, weather information updates are wholly dependent on radio communications. These radio communications are, in turn, dependent on the aircraft being within range of the broadcast antenna. Moreover, broadcast weather and terminal information update rates differ depending upon the system available. For other aircraft, a digital data link available through satellite eliminates the problem of being in range of a broadcast station entirely. Information updates can be achieved at a much higher rate and information availability is achieved much earlier in the flight than is otherwise possible.

ATC communications delivery also affects pilot cognitive activity. Communications traffic from ATC to the aircraft can occur at any time, including at times in which the pilot or aircrew are preoccupied with other tasks. This significantly impacts aircrew workload and planning activities as well as the manner in which ATC resources are used (Lee, 2003). ATC communications simulation needs to deliver ATC communications at rates comparable to those of the operational environment in order to assure the same level of monitoring activity from pilots. Decision-making will also be affected by the timing of information received from ATC. Information on delays due to airport traffic can affect the diversion decision and information on weather can affect routing decisions.

Current regulatory requirements reflect a widely-held and relatively narrow view of the aircraft simulator as simply a ground-based surrogate of the aircraft. Little consideration has been given to the role of the environment within which the aircraft operates. In order to address the importance of cognitive fidelity in simulator design, greater importance must be placed on understanding the higher-level cognitive tasks that pilots must perform. Recently, commercial airline pilot training has recognized the importance of these tasks and created training regimens which stress the importance of LOS training. In commercial airline and military LOS training simulation, the goal is to replicate the line operations environment in every detail possible in the belief that this will permit more valid preparation of the aircrew for the line operations environment. By replicating the task environment so closely, the LOS environment typically produces a high level of cognitive fidelity. Such a task environment requires the aircrew to allocate attention, solve problems, make decisions, communicate intentions, and retain information in a manner comparable to that of the operational aircraft environment. The developers of future flight simulation environments need to incorporate this new use of flight simulators in their plans for new simulation systems. The use of the flight simulator as a place where line-oriented skills are developed and evaluated requires not only a thorough understanding of pilot cognitive tasks but also of the increasingly complex information environment in which aircraft will operate.

Simulating Advanced Information Systems

In order to achieve a high degree of cognitive fidelity, it is necessary to replicate the *information* environment within which pilots and aircrews operate. Replicating an information environment means reproducing not only information content, but also the information transmission and display characteristics of the operational aircraft. Pilots receive and process information from a variety of sources and these sources are continually competing for attention. Apart from the external visual scene, control system feedback, aircraft motion and sound effects discussed earlier, information is received from aircraft instruments, from electronic and paper charts and documents, from other crewmembers and from ground-based communications. In modern aircraft operations, avionics and ground-based information sources, including radio and digital data link communications, are more important to aircraft operations now than ever before.

A vast array of instrumentation is now available in the cockpit which needs to be accurately simulated, not simply to enhance physical fidelity or realism, but also to allow the pilot to develop the attention allocation strategies and workload management skills so vital to the operation of modern aircraft. In order to achieve the level of desired cognitive fidelity in a flight simulator, it is necessary to provide the demands on pilot information processing capabilities that are comparable to those of the aircraft.

Information systems which support the operation of the aircraft in the airspace also need to behave in a manner comparable to the operational aircraft that exist *outside* of the aircraft cockpit. This is particularly true of information systems that depend on ground-based navigation, systems which transmit atmospheric and weather information, as well as information on airborne and ground-based traffic. For example, radio-based navigation equipment is limited in range and can be blocked by terrain. Aircraft weather radars can be rendered ineffective by ground clutter or may lack effective penetrating power. GPS navigation is subject to failures due to satellite transmission problems and is subject to accuracy problems under certain conditions. Replicating the characteristics of these systems in the flight simulator environments allows the pilot to develop strategies and skills to deal with them in the operational environment.

As discussed in an earlier chapter, communications simulation remains among the least developed of all simulation component technologies. Communications simulation and especially ATC communications simulation are essential if a flight simulator is to achieve a high degree of cognitive fidelity. As with other information systems, the content, transmission and display characteristics of all forms of radio communications simulation have to correspond to the environment within which the aircraft operates. For ATC communications, this means that the frequency of transmissions, their content, as well as their clarity and intelligibility need to replicate the relevant operational communications environment. The cognitive processing of communications is affected by all of the aforementioned factors, so attention to these details during the simulation design process is essential.

Summary

Modern flight simulators are capable of providing significant levels of physical realism for pilot training and evaluation. With the increasingly automated flight operations and the need to improve crew and information resource management skills, simulator task environments must provide the conditions in which cognitive skills such as decision-making, problem solving and workload management can be learned and evaluated. Meeting the challenge of cognitive fidelity in flight simulation requires a clear understanding of the cognitive, as well as perceptual and perceptual-motor tasks required of pilots. Additionally, the advances in information systems technology require that flight simulations incorporate these systems at a high level of functional fidelity. Future flight simulator designs will need to provide an environment in which line-oriented, cognitive skills can be trained effectively.

Chapter 6

Flight Simulators in Pilot Training and Evaluation

The modern flight simulator of today has origins that extend back over 70 years, to the development of Edwin Link's trainer of 1929. Link's trainer had a basic set of instruments, no visual display and a primitive motion platform. The trainer had difficulty finding a market as aviation was in its infancy and the trainer's design was centered on instrument flight training at a time when instrument flight was, at best, a limited and very risky activity. Link's trainer and others of its kind did not see great use until the years immediately preceding the Second World War. In these years, the need for rapidly training very large numbers of pilots became a necessity. Increasingly, ground-based training was sought as a means of supplementing aircraft training in military training programs.

Since the end of that war, flight simulation has been driven largely, though not exclusively, by the need to train and evaluate pilots and aircrews efficiently, safely, and effectively. As the cost of purchasing and operating a flight simulator is only a fraction of the cost of the aircraft it represents, simulator training became a logical substitute for the training of pilots in real aircraft. Moreover, many maneuvers, such as engine failures, cross-wind landings and others, could be trained in the simulator without exposing the pilot and instructor to the hazards of these maneuvers in real aircraft. Simulators also evolved in efficiency by developing the capability of positioning and repositioning the simulated aircraft at any point in space and in any configuration desired. This meant that a given maneuver could be practiced much more frequently in a given training hour than was possible in the aircraft. For many tasks, the simulator became a much more efficient and safe training platform than the aircraft.

For these reasons, the modern aircraft simulator has become the device of choice for training and evaluating pilots and aircrews. Indeed, commercial airlines now conduct all their training and most of their pilot evaluations in flight simulators. Military flight training, although less driven by the economics of flight training than the airlines, also depends heavily on flight simulators for training. Air combat training in simulators, for example, is now commonplace. Corporate turboprop and jet aircraft operations are also increasingly dependent on flight simulators for pilot training. Recreational and other non-business general aviation, however, still use the aircraft as the primary training device, although significant portions of instrument flight training are conducted in flight training devices for this segment of the aviation community.

The reasons for the differences in simulator use in different segments of the aviation industry are varied.   The cost of a flight simulator, which includes purchase, operation and maintenance, facilities and personnel training, can be substantial.   High fidelity flight simulators can cost millions of US dollars to purchase and hundreds of thousands each year to operate and maintain.   Even lower fidelity flight training devices without a motion platform or visual display can cost hundreds of thousands of dollars.   Additionally, for larger flight simulators with full platform motion systems, special multi-story facilities with reinforced foundations and a host of safety features must be built to house them.

Simulator Training Effectiveness

Since the flight simulator, unlike the aircraft, serves only to train and evaluate pilots, the primary return on the simulator investment is achieved by reducing or eliminating the use of the aircraft as a training device.   The operating cost of the aircraft used for training will drive the decision to purchase and use simulators for training more than any other factor.

So important is the reduction of aircraft operations as a justification for simulator use that a formula was developed for calculating the savings.   This formula, the Transfer Effectiveness Ratio (TER) calculates the effectiveness of the simulator by measuring the amount of aircraft training time use of the simulator saves.   The following formula is used to calculate *TER* (Roscoe, 1980):

$$TER = \frac{A - AS}{S}$$

where:

A = aircraft training time without simulator training
AS = aircraft training time with simulator training
S = simulator training time

For example, if an average of 20 hrs is needed to train a group of pilots to make a precision instrument approach using only the aircraft and another group of pilots trained in a simulator for 5 hrs subsequently takes only 10 hrs to achieve the same level of proficiency in the aircraft, the TER is 2.0.   In other words, for every 1.0 hour spent in the simulator, 2.0 hrs are saved in aircraft training time.   A negative TER would mean that there is negative training transfer, i.e., the simulator training results in *more* aircraft training time rather than less.

The TER properly used is the most reliable and valid means of measuring simulator training effectiveness.   However, care must be taken in how the measure is derived.   For example, if *total* simulator time is used rather than training time on specific maneuvers, the simulator's effectiveness can be artificially inflated due to the ability of the instructor to rapidly reposition and reconfigure the simulated aircraft. These simulator features allow the instructor to provide the trainee pilot with more opportunities to learn a given task in a given training hour than is possible in the aircraft. The flight simulator therefore has a built-in advantage over the aircraft in how the training time is actually used. In terms of time spent on training a *particular task*, the simulator could actually be *less* effective than the aircraft since it may take several more repetitions of the task than it would in the aircraft. A better use of TER includes one calculation for total simulator and aircraft time and a second calculation for time on task (TOT) in the simulator and the aircraft. This allows an assessment of the effectiveness of the simulator as an instructional device independently from the effectiveness of the simulator as an aircraft surrogate in training specific skills.

While the TER is an important measure of simulator training effectiveness, conducting the necessary field study to obtain the transfer data required by the TER formula is difficult due to the high cost and substantial logistical problems involved in conducting transfer-of-training studies. As a consequence, there have been only a few dozen transfer-of-training studies conducted in the entire history of flight simulation development. Due to the dearth of available studies, data necessary to calculate TERs for many flight simulator devices in use today do not exist. For example, no transfer-of-training study has ever been conducted on the FFS systems commonly used for commercial airline training. As a consequence, no TERs for these devices are available.

Despite the obstacles to data collection, some transfer-of-training studies evaluating flight simulator devices have been conducted. A review of these studies is merited as they provide some quantitative data on the actual training effectiveness of flight simulators and therefore on the expected savings in aircraft time that can be expected. Wherever possible, the actual TERs will be provided if available from the original study.

Simulators Used for Primary Training

Primary or *ab initio* training represents the first step in the process of pilot development. Primary training introduces the student pilot to the fundamentals of flight by reference to visual cues from the scene outside of the cockpit. While use of basic aircraft instruments such as airspeed, altitude and heading indicators is important, the use of visual cues for basic aircraft control is essential. This means that the training device needs to provide a display of the external visual scene of sufficient perceived fidelity to allow for basic visual flight skills, such as approach and landing to be trained. As quality visual scene simulation is a relatively recent phenomenon, the training effectiveness of devices designed to train basic visual flight skills have only been evaluated relatively recently.

One of the earliest transfer-of-training evaluations of a primary training device was conducted at the University of Illinois (Lintern, Roscoe, Koonce, and Segal,

1990). Using a fixed-base, small aircraft trainer with a single channel, color visual display (26.3 deg horizontal × 19.5 deg vertical), student pilots with no prior flying experience were trained to proficiency first in the simulator and then in the aircraft. The results of the study showed average savings of 9.8 landings per student for those trained in the simulator when compared to those with no simulator training. Aircraft time saved was 1.5 hrs for every 2.0 hrs in the simulator for a TER of 0.75.

A more recent study using the popular Microsoft Flight Simulator (v. 4.0) personal computer (PC) game program also found positive training transfer for visual flight skills (Dennis and Harris, 1998). The program incorporated a single channel, color visual scene display (45 deg horizontal and 28 deg vertical). Students were trained in straight and level and coordinated turns in the device for 1 hr. Following training, the student pilots were evaluated in the aircraft. As with the previous study, the target aircraft was a small, single-engine, training aircraft. Students who had received training with the simulation device performed significantly better in the aircraft than students who had not received training in the device.

These two studies demonstrate that devices of relatively low physical fidelity can provide effective primary training, provided a minimum level of perceptual fidelity is maintained. A limited FOV visual display is adequate for the basic training of some key visual flight skills including landing skills if the display presents the essential visual cues necessary to support the task. A basic level of instrumentation is also needed, but the instrumentation need not be a perfect physical replica of those in the aircraft. The Microsoft Flight Simulator displayed instrumentation at only a fraction of normal size.

The study by Dennis and Harris (1998) also noted, however, that the primary control system (e.g., yoke and rudders) used in the Microsoft Flight Simulator were no more effective than simple keyboard controls. The authors concluded that the perceptual-motor skills normally associated with visual flight could not be effectively trained on such a device. However, the tasks were relatively simple and did not include takeoffs and landings so the conclusions drawn must necessarily be limited.

Earlier transfer of training studies conducted by the US Military also supports the role of ground-based training for primary visual flight skills. Conducted in the late 1970s, the device used had relatively primitive (by today's standards) CGI visual scene simulation, but did have high fidelity instrumentation and control landing (Martin and Cataneo, 1980). The simulator, designed to support T-37 jet training, proved effective in training takeoff maneuvers under both day and night conditions. A later study also demonstrated that varying the size of visual display FOV did not alter the device's training effectiveness (Nataupsky, 1979). Both studies once again demonstrated that primary training can be conducted in simulators without the need for expensive, high fidelity visual systems.

To date, only a few transfer-of-training studies have been conducted which have evaluated the efficacy of flight simulators for primary flight training. Partly, this is due to a strong bias within the training community towards the use of aircraft, not simulators, for primary training. As most primary training is conducted in small, piston-driven aircraft, the operational cost of this type of training is much less an issue than if larger aircraft are used. The use of simulators for primary training has also been hampered by the lack of high fidelity visual scene simulation, though this reluctance has probably been largely unjustified. It is likely that the combination of escalating cost of aircraft operations and the rapid developments in visual scene simulation technology will result in greater use of flight simulators for primary training in both civilian and military sectors.

Simulators Used for Instrument Training

Unlike primary training, the aviation community has largely embraced the use of simulators for instrument training. As instrument flight, by definition, relies on the pilot's use of aircraft instruments for flight control and navigation, the issue of visual scene simulation fidelity is moot. Many instrument flight trainers, in fact, have no visual display capability at all. Additionally, many of these devices have relatively low fidelity control loading and control feel as instrument training is largely the training of *procedural*, rather than perceptual-motor skills.

Additionally, instrument trainers do not normally have motion platforms. This makes them more affordable to operate and maintain and to use without the requirement of special facilities. Flight Training Devices (FTDs) and Personal Computer Aviation Training Devices (PCATDs) are categories developed by regulatory agencies, such as the FAA, to support instrument training and all do so without whole body motion cueing.

Some of the earliest transfers of training studies were conducted on instrument training device predecessors of the FTD. The general aviation trainer (GAT) developed by the Link simulation company is an example. Provenmire and Roscoe (1971) conducted a transfer-of-training study that evaluated the transfer-of-training effectiveness of the 6AT-I trainer to a single-engine aircraft. The 6AT-I had no visual system, yet delivered TERs averaging 1.0. This was in spite of the fact that primary pilot training, not simply instrument training, was the goal of the program. For this study, the trainer was essentially equivalent to the aircraft for primary flight training despite the absence of visual scene simulation capability.

A similar finding of positive transfer of instrument training to contact flight has been reported by Pfeiffer, Morey and Butrimas (1991). In this study, pilots transitioning from a turbo-prop training aircraft to a centerline thrust, twin turbojet trainer were trained in a turbojet trainer prior to transition; the trainer had no visual display, but did have a 6-DOF motion platform. TERs for aircraft visual flight maneuvers and instrument flight maneuvers were nearly identical (0.43 vs. 0.44). The study also showed that the particular mix of ground-based and aircraft training had no effect on the study outcome.

The positive transfer of training from instrument to contact flight is explained, in part, by the fact that only a few visual flight maneuvers depend *primarily* on visual scene cues while most maneuvers share skill components with instrument flight. For example, final approach and landing, and particularly landing flare maneuvers depend essentially on the proper use of visual cues by the flight. But most basic flight maneuvers such as heading, airspeed, altitude, turn, rate of climb and many others can be done with reference primarily to aircraft instruments. The evidence from transfer-of-training studies with instrument simulators suggests that significant training time in the aircraft could be saved by use of instrument flight training devices in the primary flight training curriculum.

The above studies are but a few examples of the successful application of simulator technology to the training of pilots. Instrument training seems especially well-suited to these ground-based training devices. Since the training is not dependent on complex visual scene simulation or motion cueing systems, it is much less dependent on advancements in computer technology than other forms of flight training. The nature of instrument training, with its emphasis on procedural skills, also makes it a viable candidate for the use of lower cost, desktop training device, such as the desktop PC. Unlike the 6AT-I or other similar devices now classified as FTDs, the PC-based devices provide a very restricted instrument display surface, very low fidelity control devices and either no visual displays or a display with very limited FOV. Additionally, interaction with instruments often requires the use of a pointing device such as a computer mouse or the use of keyboard inputs. The use of radio and other instruments sometimes requires that they overlay or cover flight instruments due to the lack of display real estate on desktop monitors.

The very low cost of PC-based training and the PC's potential for easy access by pilots makes PC-based instrument training very attractive. A study of the transfer-of-training effects of PC-based training of limited instrument maneuvers (straight and level flight and coordinated turns) showed clear benefits of such a device (Ortiz, 1994). When evaluated in a single-engine, propeller-driven training aircraft, a large transfer effect was shown from the device to the aircraft. One hour in the device saved up to 29 min of aircraft time for a TER of 0.48.

A more recent study examined the training value of a similar PC-based device (Taylor, Lintern, Holin, Talleur, Emanuel, and Phillips, 1999). In this study, an FAA-approved PCATD device was evaluated as a part of a university instrument training course. A wide range of instrument flight tasks were evaluated in this study including complex holding patterns and instrument approaches. The training aircraft was a single-engine, propeller driven aircraft.

The TERs were much less impressive in this study than the study by Ortiz (1994). TERs ranged from a low of -0.11 to a high of 0.39. The only reliable differences were found for ILS approach training (0.28), NDB holds (0.16). NDB approaches (0.39), LOC BC holds (0.30) and VOR tracking (0.17). The overall savings in aircraft time by using the PCATD for instrument training was 3.9 hrs. In this study, only 1.5 flight hrs were saved for ten hrs of PCATD training for an overall TER of 0.15. This TER is far less than those reported above for either FTDs or PC-based devices and less than the average TER of 0.45 reported by Orlansky and String (1977) for military flight simulators.

Taylor, et al. (1999) suggests that the low overall TERs of their study is at least partly due to the inefficient use of the PCATD required by the organization of the training curriculum. Due to the division of the curriculum into two separate classes, much PCATD time was devoted to reviewing previously learned tasks. This contributed to the low TERs found in the study.

The existing data, although limited, provide some confirmation that relatively low cost devices such as PCATDs can be effective in instrument training. Exactly how effective they are remains to be determined by future studies, however. An important and, as yet, unanswered questions is whether PCATDS are as effective as FTDs for instrument training. If so, major cost savings and far greater access to a valuable instrument training tool could be achieved for the aviation community.

Specialized Training

Studies assessing simulator transfer of training have also been conducted in more specialized areas of aviation. The military aviation community, which depends heavily on flight simulators for training, has conducted a number of training transfer studies. A complete review of all of these studies is beyond the scope of this book. However, some of the studies reveal the particular challenges of military aviation flight simulation and are included here for that purpose.

Military aircraft serve in the primary roles of air-to-air combat, ground attack, air transport, surveillance, and search and rescue. These and other missions require a large variety of fixed and rotary wing aircraft with unique and demanding mission requirements. Flight simulators serve in the military not only as primary and instrument training devices, but as weapons and mission training tools. In the latter case, military flight simulators not only reduce the hazards associated with pilot training, but provide an opportunity to train combat mission skills which are very difficult or impossible to train in any other way. The advanced skill training requirements of military aviation training place a far greater demand on simulation technology than is the case for civilian aviation. This demand dramatically increases the potential cost of military flight simulators and consequently requires greater attention to the actual training benefits of the devices.

A recent review of US Air Force and Navy studies emulating combat simulator training effectiveness was conducted by Bell and Waag (1998). For the air-to-surface attack role, simulator training provided consistently superior pilot performance in conventional weapons delivery when compared to pilots who had received no simulator training. This was the case across a wide variety of aircraft types. The studies also revealed a significant increase in the probability of aircraft survivability for those pilots trained in combat simulations of high-threat environments.

Bell and Waag (1998) also reviewed training transfer studies of air-to-air combat simulation training effectiveness. Only a few such studies have ever been conducted, but all save one showed positive transfer-of-training effects. The authors suggest that the use of subjective rating data in this one study lacked sensitivity sufficient to detect training differences.

While the authors did not report TERs for the various transfer studies, any reliable effect of simulation training is likely to justify the cost of these combat simulations given the extremely high cost of full mission field simulations, the hazards involved in field exercises, and the limited range of threats these exercises can deliver.

Simulator Component Effectiveness

Transfer-of-training effectiveness has also been assessed for key components of flight simulators, such as visual scene simulation and whole body motion cueing systems. The rationale for assessing these components separately is typically one of cost-effectiveness. If the amount of training transfer is comparable for a fixed vs. motion platform, for example, significant cost savings could be achieved by eliminating the motion platform from the design. For this reason, such studies have been limited to those components that could have a significant impact on the purchase or operational cost of the device.

Motion Platform Systems

Few issues in flight simulation design have raised as much controversy as that of platform motion. While anecdotal evidence as well as empirical studies (Burk-Cohen, Saja, and Longridge, 1998) suggests strong support among pilots for the provision of platform motion in simulators, there is as yet no data that supports the assertion that motion platforms actually improve the training effectiveness of simulators. A review of six transfer-of-training studies by Martin (1981) conducted on primary and advanced (air-to-air) jet training simulators found no reliable increase in transfer of training when motion platform cueing was added to the flight simulator. Another study of motion platform effectiveness on the training effectiveness of a four-engine, turboprop transport aircraft (a P-3 Orion) also failed to show any added benefit (Ryan, 1978). Recent reviews of motion platform effectiveness literature by Hays (1992) and by Burk-Cohen, et al (1998) have come to the same conclusion. There is as yet no reliable evidence that the training effectiveness of a flight simulator is improved when a motion platform is added.

Several possible explanations for the lack of scientific evidence in support of motion platform training effectiveness have been put forward. These include deficiencies in the platform systems used, poor study methodology, and the use of measures of pilot performance that may have been insensitive to differences in motion cueing during training. The striking absence of evidence over the past 20 years of study, however, suggests that motion cueing is probably of little or no importance and that the inability to find reliable improvements in training is due to the fact that motion cueing systems used in modern flight systems are simply not effective. This is at least partly due to the limitations of the systems themselves, to the fact that very few flight maneuvers require a pilot's attention to motion cues, and that the small effects of motion cueing that might exist are masked by a variety of other phenomena when pilot performance is assessed in the target aircraft.

Despite the lack of supporting evidence, motion cueing systems are still required by regulatory agencies in the civilian sector for use in FFS systems, although they are less likely to be found in military simulators where motion cues from large FOV visual systems have been preferred. It is likely that motion platform systems will be required in some civilian flight simulators for the foreseeable future owing to the strong preference for such systems by the pilot community. The question will then be whether some improvements to the basic design of these systems might be possible that could directly benefit pilot training.

Visual Scene Simulation

While much less controversial than motion platform systems, visual scene simulation can also impose significant costs on flight simulator development. These costs are increased substantially by requirements for high effective resolution, large display FOV, and high scene details. As with motion platform cueing, there is typically a strong desire on the part of the simulator user community for more visual scene simulation capability, rather than less. While task analyses can identify most of the requirements of visual scene simulation for given flight task, not all tasks are amenable to this type of analysis. Tasks that have become highly automated by the pilot with repeated practice are difficult to decompose in a manner that allows identification of relevant visual cues. Many continuous control tasks such as the landing flare and low level flight fall into this category.

Visual scene simulation is most important in areas where direct aircraft control is dependent on the visual scene. In systems where visual frame update rates are too slow, the delayed control feedback to the pilot results in significant control problems. This is especially true where more sensitive pilot control inputs are required, such as during the landing flare (Bradley, 1995). Visual scenes which have wide FOVs that are incapable of presenting sufficient image detail will likely be ineffective in teaching basic visual flight control skills. The ability to detect altitude change appears to be dependent on the number of objects displayed, with high object density resulting in more accurate altitude change judgments (Kleiss and Hubbard, 1993). Transfer of training from landing skills training on simulators with low-detail visual scene displays has also been shown to be poor (Lintern and Koonce, 1992). Inadequate visual FOVs can also result in the development of compensatory visual behaviors as demonstrated in a study of C-130 aircrew performance in low level flight maneuvers (Dixon, Martin, Rojas, and Hubbard, 1990). Limited visual FOV in the device resulted in a significant change in out-the-window visual scanning behavior by the pilots.

Comparisons have also been made between various types of visual display systems with respect to their effectiveness in pilot training. In a study comparing visual dome systems commonly used in air-to-air combat training with helmet-mounted displays (HMDs), the dome display was found to be the superior visual system when visual target performance and user workload were measured (Hettinger, Todd, and Hass, 1996).

Summary

The development of flight simulators has been driven largely by the role they play in the training and evaluation of pilots. This naturally raises the question as to how effective they are in this role, whether certain components of flight simulators are worth the cost, and where improvements might be made. Strong evidence for the training effectiveness of flight simulators including PC-based training devices exists. Evidence for the effectiveness of certain components, such as motion platforms, was not found. Visual scene simulation effectiveness is largely dependent on the presence or absence of specific image details and the task to be trained. Object density and other scene details as well as visual display update rate were found to be especially important for the training of visual flight control tasks.

Chapter 7
Simulator Fidelity and Training Effectiveness

Introduction

Flight simulator design has been driven since the early part of the last century by a design and engineering concept that equates the value of the device in direct proportion to its ability to physically replicate the aircraft's controls and displays, handling qualities and other features. This includes not only the physical layout and appearance of the aircraft cockpit, but the function of individual controls and displays. When simulator engineers speak of fidelity, it is generally the physical replication or *physical* fidelity of the simulator design to which they refer. Physical fidelity is extended to visual scene simulation, motion accelerations, and all other aspects of flight simulator as a means of identifying the criteria by which a particular flight simulator design should be measured. For the design engineer, the closer the design is to the ideal of full or complete physical fidelity in all its dimensions, the better the simulator design will be and the more effective the flight simulator will serve as an aircraft surrogate. This stress on physical fidelity can be seen in the terms used by the FAA in categories of flight simulators used for training. For example, only those devices which have visual scene simulation and motion cueing can be called 'flight simulators', all others must be called 'flight training devices' or 'aviation training devices'.

The prevalence of physical fidelity as a criterion of simulator design, whether it is in instrument layout and function, handling characteristics, or any other design component, is understandable. If the *intent* of a flight simulator designer is to create a replica of the aircraft flight environment in a ground-based device, then it is reasonable to attempt to use a physical reproduction of that environment as the design goal; the closer the reproduction, the better the simulation. Unfortunately, this measure of simulator design suffers from two major flaws. First, the primary purpose of a flight simulator is not to replicate the physical flight environment, but to create the *experience* of a flight environment for a pilot or aircrew. The ultimate measure of simulator fidelity is the ability of the simulator to elicit pilot or aircrew *behavior* that is indistinguishable from behavior that occurs in the operational aircraft under similar circumstances. This behavior includes not only the manner in which the simulated aircraft is flown, but also the *cognitive* behavior of the pilot as expressed in the way in which the pilot makes plans, decisions and solves problems. Physical fidelity is only important in the flight simulator to the extent that it provides the conditions under which a high degree of this psychological fidelity can be achieved.

The second major flaw in the argument for physical fidelity as a design goal is that it allows for no means by which design limits can be logically justified. If the flight simulator motion platform cannot meet the criterion of physical fidelity with the aircraft because of prohibitive costs, for example, the designer is left with no logical means of determine what the effects of less than full physical fidelity will be. There is no rational basis to assume that simulating 60 percent of aircraft motion is more realistic or provides better training than simulating only 40 percent, for example. Finally, in the absence of any references to the limitations and capabilities of the human pilot, simulators designed to maximize physical fidelity are not cost effective. That is because design limits will be imposed on the simulator based on criteria that are largely irrelevant to the purpose for which the simulator has been built. The most common limits on flight simulator design criteria are cost, limits in technology and the physical limits imposed by the training or research facilities. These are not criteria that are relevant to the training effectiveness of the design.

As attractive as the goal of maximum physical fidelity might be to the flight simulator engineer, it is not a means by which cost-effective simulation can be achieved. Yet physical fidelity remains compelling because it provides a clear and unambiguous reference point for a design. Reproduction of aircraft instrumentation, control responsiveness, and other physical characteristics of aircraft operations have a ready-made benchmark in the aircraft itself. There is little ambiguity in a design goal that is based on simple reproduction of an object or system. Physical fidelity of a flight simulator is also used as a means of convincing the simulator purchaser of the value of the device since it is easier to accept that an aircraft simulator that appears and functions more like that of the actual aircraft will be more valuable than a simulator that does not.

However, problems exist for design engineers in those areas of flight simulation where the human operator plays a significant role in the processing of information provided by the aircraft. In the areas of simulator control loading, visual scene simulation, platform motion, and communications simulation, for example, the human pilot processes these inputs within the framework of his or her own sensory capabilities, knowledge and experience. In this case, the effectiveness of the simulator design is not a simple matter of reproducing the appearance or functionality of systems, but of the interaction between the pilot and the simulation system. For the trained pilot, maneuvering motion cues, no matter how faithful they are to the real aircraft, may be of no consequence if they do not provide useful information in determining the state of the aircraft or the state of the environment around the aircraft. The provision of such motion cues in a simulator may in some way enhance user acceptance of the device, but it may serve no other purpose. Conversely, a device such as a PCATD which is very low in physical fidelity can provide valuable training because it provides the necessary information and pilot response capability essential for the performance of a particular task.

Devices that are objectively low in physical fidelity are often found to be valuable in meeting research and training goals because they are high in psychological or perceived fidelity.

Perceived Fidelity in Simulator Design

An alternative to the traditional design approach, which relies on replication of the aircraft's physical or functional characteristics, is clearly desirable.   It is particularly desirable if flight simulators are to achieve a high degree of cost-effectiveness in pilot training. An alternative design approach is needed which is based upon the fundamental assumption that flight simulators are intended to re-create the pilot or aircrew operational aircraft flight experience in a ground-based device. Those design elements of the simulator that do not support this goal might have other uses, including market appeal, but they do not contribute in fulfilling the essential design goal.

A view of simulator fidelity which stresses the importance of experienced or perceived fidelity as the primary design goal will face resistance from traditional engineering disciplines since it is removed from the familiar arena of traditional engineering disciplines. Traditional engineering disciplines have often found it difficult to deal with systems designs where human operators play a significant role. This is partly because of the difficulty of quantifying human operating characteristics. Engineering models of human perceptual and cognitive functions equivalent to those typically found in aeronautical engineering, for example, are in short supply. As a consequence, the data necessary to support design decisions based on perceived fidelity are often not available. However, the reliance on the classic engineering approach to addressing many of the essential deign issues in perceived fidelity is probably overstated. It is possible to achieve this design goal with much less than the engineering rigor required of other methods. Many of these methods are well known to individuals in disciplines that focus on human-systems integration, such as industrial and human factors engineering. Some of the key elements of human-systems integration engineering are briefly described in the following sections.

User Profiling

Humans vary significantly in their limitations and capabilities to perform tasks. User profiling is a means by which the system designer can identify those characteristics of the end-user that are most relevant to the design decisions that must be made.   User-profiling is a means of identifying the device user population(s) and their specific capabilities and limitations. The typical user populations for a flight training device would include the primary (pilot or aircrew) population and a secondary population of system operators such as instructors or research scientists. The primary user population of pilots will have a range of basic perceptual, perceptual-motor and cognitive capabilities. The designer of a flight simulator used to train beginning or *ab initio* pilots can safely assume that the primary user population has normal vision, hearing, vestibular and cognitive functions since these are required of all pilots. However, this user population will likely differ dramatically from other pilot populations in their level of manual flight control skill, procedural knowledge, and decision-making ability.

It is also true that the secondary user populations may also differ in the way in which each uses the simulator. Instructors and evaluators make up the typical secondary user population of flight simulators used for training. These secondary users will likely have considerable expertise in the particular aircraft or class of aircraft being simulated. They may, however, lack skill in the actual use of the simulator as a training device. Instructional use of the device for teaching basic skills requires use of simulator features such as simulator freeze and re-positioning because much more training time is spent in diagnosing problem areas and repeating task scenarios. When used as an evaluation tool for experienced pilots, such as in LOFT sessions, much more stress will be placed on the ability of the simulator to reproduce conditions where incident and accidents are likely, such as instrument and engine failure or hazardous weather conditions.

Understanding the Piloting Tasks

The next step towards perceived fidelity in flight simulator design is to identify the piloting tasks that the simulator must be able to support. To isolate the information, control inputs and other elements of tasks pilots perform requires a detailed and rigorous task analysis. There are dozens of forms of task analyses, the details of which are beyond the scope of this book (see Kirwan and Ainsworth, 1992). However, most task analyses share common components. First, a definition of *task goals* or objectives is needed. Defining task goals is much easier in aviation than other fields because piloting tasks have generally clearly defined goals and sub-goals and well-defined procedures or steps necessary to achieve these goals. The means by which goals and sub-goals achieved are the *task elements*. It is these task elements that contain the information needed to guide the simulator design. Aircraft state parameters such as airspeed are critical to many flight tasks. Airspeed *information* can come from a variety of sources including instrumentation, the visual scene outside the cockpit window, wind noise and related sound effects, and others. The pilot's *response* to this and other information is another part of the task element analysis. The responses of the pilot may take a variety of forms to specific information depending on the task goals. Achieving a targeted airspeed during approach generally requires visually monitoring of an airspeed instrument, changes in power settings, and changes in flap and landing gear configuration as a means of achieving and maintaining the desired airspeed. For the trained and experienced pilot, there is a known *relationship* between information and a particular response. For simulators used in basic flight training, the development of this relationship in the trainee pilot must be supported by the flight simulator. For the experienced pilot who is undergoing an evaluation in the device, the simulator as an evaluation device needs to be able to provide the necessary cues to elicit the correct pilot behavior for the prevailing conditions if it is to be effective in measuring pilot proficiency.

Task Environment

Understanding in detail the characteristics of the user population and the goals and elements of piloting tasks are essential to the development of flight simulators with high perceived fidelity. In addition, simulator designers need an understanding of the task environment within which pilots and their aircraft operate. The task environment includes all those attributes that might directly or indirectly affect the pilot's operation of the aircraft. Such an understanding is needed in order to assure that the design of the flight simulator incorporates attributes of the task environment that might impact perceived fidelity.

Simulation of the task environment for flight operations includes a clear understanding of the complex conditions under which flight operations take place. Such operations often take place within a complex, controlled airspace in which aircraft speed or direction are dictated by someone on the ground. Such operations always include references by the pilot to clearances, flight plans, maps and charts. They may include the requirement for continuous monitoring of radio frequencies for ATC directives, the use of radio communications, and the use of a digital data link for the transmission of vital data.

The pilot's task environment is further complicated by the often unpredictable changes in weather. The weather not only makes the flying task difficult and potentially hazardous, but can have a major impact on the most elemental aspects of aircraft operations. Low level wind shear, and especially microburst wind shear, can make an aircraft extremely difficult to control, as can icing on wing and control surfaces. Icing can also lead to engine stalls in carbureted engines and engine failure in turbo jets due to ice ingestion. Rapid changes in weather can result in lowered ceilings and visibility making landings and takeoffs difficult or even unsafe. Major shifts in winds aloft can result in excessive fuel burn rates which can lead to the necessity for emergency diversions, as can the sudden onset of thunderstorms and other severe weather. Of all the attributes in a pilot's task environment, weather is perhaps the single most important attribute that the simulator designer needs to consider.

The simulator designer also needs to consider the primary purpose of the device, which is typically instructional. Simulators are typically purchased to train and evaluate pilot flying skill. Aircraft, on the other hand, are designed to carry passengers and cargo to a destination or to carry out a military mission, or for some other purpose. Generally, aircraft are, in comparison to a simulator, a poor platform for training. The instructional aspect of the flight simulator environment is, perhaps, the least developed aspect of this device. Once again, because of the focus on replicating the physical characteristics of the aircraft in simulator design, comparatively little attention, effort, or technology has been spent on this critical aspect of the simulator task environment.

Measuring Perceived Fidelity

However fidelity is defined, there must be a means by which it can be measured. Physical fidelity of a simulator is measured directly against its operationally equivalent component. If we wish to know whether the simulated aircraft performs in a manner comparable to that of the operational aircraft, the simulator's performance can be compared directly to the aircraft's performance data from documents supplied by the aircraft's manufacturer. For example, takeoff and performance data in terms of power settings, flap configuration, and so on for a standard day and temperature are readily available and can be compared with the same test data from the simulator.

Where such objective data exist on those components of the simulator that are not subject to mediating influences of the pilot, they can serve an effective role in meeting the goal of fidelity. In the case of simulated aircraft performance, the objective is not to elicit or create a particular pilot experience, but simply to match with minimal error the aircraft's behavior under specific conditions. Similarly, objective criteria exist for radio transmission reception ranges, instrument performance and acceptable instrument error, engine power output under various conditions, aircraft and engine performance effects in icing conditions, electric and hydraulic system behavior, and a wide variety of other aircraft system performance parameters. In short, the simulator designer has available a large amount of data which can be used to assess the physical or functional fidelity of the aircraft and its subsystems, as well as systems that it may depend on within its operating environment, such as navigation and communications systems. In assessing the physical fidelity of a simulator, the definition of both the criteria and the measurement of fidelity are relatively straightforward. Published engineering criteria on the performance of aircraft, aircraft component systems, and airspace or other environment systems are readily available. The performance parameters for these systems are generally well-known, although some, such as aerodynamic modeling data for a specific aircraft, are proprietary and may require a license to use. Problems in meeting physical fidelity requirements on the part of the simulator designer are more often related to limitations in software and hardware engineering. For example, limits in computer processing power can significantly limit the ability of the simulator to solve the relevant aerodynamic or aircraft performance model equations that underlies accurate simulation of aircraft behavior. The use of analog instruments driven by small direct current (DC) motors or DC metering systems are likely to introduce instrument errors simply due to inherent problems in providing the precise voltages to these instruments consistently and at the precise time needed. Improved engineering and design efforts in these areas have been able to manage the problems effectively, though the potential for errors remains. The problem of fidelity in simulator design is now, not primarily with issues of physical or functional realism, but with fidelity as perceived by the pilot or aircrew.

If the true goal of flight simulation is in creating a flight task experience for the simulator pilot that is indistinguishable from that of the operational aircraft, then the simulator designer must go beyond physical or functional fidelity and meet the more demanding and difficult requirements of perceived fidelity. While the criteria of physical fidelity are, for the most part, readily available, those of perceived fidelity are much less so. Even though there exists a substantial amount of data on human limitations and capabilities, the problem of perceived fidelity is really a problem of human-system or, to be more specific, pilot-simulator integration. The designer needs to be able to create a flight simulator which provides the pilot with a perceptual, perceptual-motor, and cognitive environment so closely resembling the operational aircraft that the pilot cannot readily discriminate one environment from the other. The ultimate goal and the measure of ultimate success for the flight simulator designer may be what the virtual environment community calls 'immersion'. Immersion is the sense the pilot has of being in the actual aircraft and being in the actual operational environment. When the pilot is immersed in the simulation, there is no conscious distinction between the simulator and the operational aircraft. Such simulation fidelity is capable of eliciting behavior from the pilot or aircrew that is also likely to be of the greatest interest to an instructor or a researcher, i.e., behavior which would have been elicited under similar circumstances in the actual aircraft. There exist subjective measures of immersion in the virtual environment literature (Witmer and Singer, 1998). There also exist measurements which can be used to analyze pilot behavior in the simulator and compare that behavior with that of the aircraft. Examples where perceived fidelity could be measured in flight simulators or has been measured in the past are described in the following sections.

Handling Qualities

Perhaps the most fundamental characteristic of a flight simulator is its ability to reproduce the handling characteristics of the aircraft it simulates. Handling characteristics are typically experienced by the pilot through forces required to manipulate primary controls to achieve the required effect. For example, a timed coordinated turn in an aircraft is a fairly common procedure under instrument flight conditions. The forces required to perform this coordinated turn with the simulator yoke and rudder controls should be indistinguishable from the pilot forces required to execute the coordinated turn in the aircraft under comparable conditions.

Handling qualities measurement has a long history in the aircraft testing field. Early aircraft were often designed with inherently poor handling characteristics. So poorly designed were some of these aircraft that they were often lethal to all but the most skilled pilots. In order to avoid the hazards of poor handling qualities in new aircraft, a standard measurement of handling qualities and characteristics was needed during the flight testing phase. Perhaps the most famous of these and the best developed is the Cooper-Harper rating scale (Cooper and Harper, 1969). The Cooper-Harper rating is a ten-point scaling system that describes the control characteristics of the aircraft being tested. While the scale was designed for operational aircraft evaluations it can be easily adapted to the problem of comparing a particular flight simulator's handling characteristics to that of the aircraft it simulates. The terms of reference need to be altered slightly, however, to make the rating scale useful for simulator design evaluation. For example, the basis for comparison would no longer be a 'stable and controllable' platform, but a simulator of a particular aircraft type and 'compensation for control problems' would be compensation for differences between the simulator and the aircraft. Obviously the pilot ratings of simulator handling characteristics must be carried out by those pilots representative of the pilot population likely to fly a particular aircraft type. Their discriminative abilities are likely to be very different from the test pilot population.

Objective measurement criteria for the handling characteristics of simulators can also be achieved by instrumenting an aircraft to record pilot control inputs. Under comparable conditions, control input characteristics of pilots flying the simulator and the target aircraft should be indistinguishable if the simulator is providing the necessary perceived fidelity. Such measures can also be used to determine the contribution of other simulator components, such as platform motion cueing, to the fidelity of simulator handling characteristics.

Visual Scene Simulation

The visual scene simulation design requirements are partly determined by an understanding of basic human visual perception, visual acuity, and visual contrast sensitivity. Much data exist in the scientific literature that can aid the simulator designer in enhancing the perceived fidelity of visual scene simulation. Some of these data have already been discussed in a previous chapter. However, there are elements of visual scene simulation that affect pilot behavior that are less amenable to quantification. The role of texture gradients in aircraft visual flight control, such as the landing flare and low level flight, is one such element. Texture and other object characteristics contribute to complex optical flow patterns in the visual scene. The changes in these optical flow patterns affect aircraft control. They also, under some circumstances, create the illusion of self motion or vection often experienced in flight simulators.

For flying tasks involving aircraft visual control at very low altitudes, the ability to discriminate optical flow patterns in the visual scene are known to affect pilot performance (Kruk and Reagan, 1983). Visual scene simulations that need to support these types of flight tasks need to be able to provide optical flow patterns equivalent to those provided in the target aircraft environment. Note that this does not necessarily mean that the visual scene simulation must provide photo-realistic visual scenes, but rather that the scene should provide sufficient image detail at a given altitude over a given FOV so as to generate the equivalent optic flow in the simulator as is provided in the operational aircraft environment.

Details for calculating unidirectional and radial optical flow patterns in visual scene simulation can be found in Regan, Kaufman and Lincoln (1986). Using quantified methods to determine the required optical flow pattern to support the low level flight task of interest also allows the development of generic task environments which provide the necessary visual scene information without having to reproduce specific object details. For example, simulating low level flight over desert terrain might be done without the need to simulate individual terrain objects (e.g., desert foliage). There would only be a requirement to create generic visual details sufficient to generate the requisite optical flow patterns in the visual scene simulation.

Motion Cueing

A simulator's capability to provide realistic motion cueing has long been a source of controversy. Once again, the ability to calculate the physical accelerations a pilot experiences in an aircraft is not an issue. Both linear and angular accelerations for specific aircraft and aircraft categories are generally known from test data collected during aircraft development. It is the contribution that these motion accelerations make in the performance of the flight task that should determine whether they will be provided in the flight simulator or not. The ability to sense a particular level of acceleration by a pilot has been discussed earlier and the thresholds of these accelerations are known. While defining sensory thresholds aids in the definition of the minimal motion cues needed, it does not answer the question as to the contribution of motion cues to the pilot's performance of a given flight task. Only detailed analysis of the task itself can provide the answer.

The most reliable means of determining the contribution of motion cues to the pilot's ability to manually control an aircraft is to compare pilot control behavior in the aircraft with that of control behavior in a simulator capable of providing the full range of accelerations that would be available to an aircraft in a given set of maneuvers. Research simulators such as NASA's VMS or other similar devices are capable of providing motion cues unrestricted by excursion limits (such as those of the Stewart platform). A recent study of this type has been conducted by Delft University (Steurs, Mulder, and Van Paasen, 2004) using the SIMONA Research Simulator system and a Cessna Citation II. Pilot control behavior in the aircraft was compared to pilot behavior in the simulator with and without motion cueing. The study found that motion enhanced the pilot's *subjective* rating of simulator realism, but had no reliable effect on pilot control behavior. This study further emphasizes, along with an earlier study by Lee and Bussolari (1989), the important distinction between the simulator user's subjectively perceived realism and the actual *behavior* of the pilot in controlling the simulator. Perceived fidelity of motion cueing, as with other elements of pilot mediated simulator fidelity, may affect the pilot's perception of realism, but may not affect the pilot's overt behavior in operating the simulated aircraft. It is possible that a platform motion cueing system's real contribution to the training or research value of a device may only be in motivating the pilot to accept the simulator as a true surrogate of the aircraft and not in the training or evaluation of pilot flying skills.

Environment Simulation

All aircraft operate within a larger environment that affects both the aircraft and the behavior of the pilots and aircrews. In the civil aviation environment, the aircraft operates in an airspace system largely controlled by people on the ground. The aircraft will encounter a wide range of weather conditions and will eventually conflict with other aircraft in the air or on the ground. Military aircraft have the added burden of operating in threat environments consisting of hostile aircraft and ground-based air defenses.

For the civil aviation environment, the physical parameters of various weather phenomenon and their effects on aircraft are reasonably well known. Airspace environment issues, including the performance of ground-based and satellite navigational aids, communications systems performance, air traffic density, and ATC performance are also well understood. However, how these environmental factors contribute to the pilot's perception of simulator realism and to pilot workload, problem-solving and decision-making behavior are not well understood. Assessing pilot workload by using a variety of workload rating scales, such as the NASA Taskload Index (NASA TLX), may be used to determine whether pilot workload levels in the flight simulator are comparable to workload levels in the aircraft under similar circumstances. More objective measures of workload, such as heart and respiratory rate, have also been used to compare levels of workload imposed on the pilot in the simulator with those experienced in the aircraft. Studies comparing these physiological indices of pilot workload in the simulator and the aircraft have been done for large transport aircraft (Jorna, 1993) and fighter aircraft (Magnusson, 2002). It is notable that, in all of these studies, physiological reactions to flight tasks were the same in the simulator and in the aircraft.

Measurements of the differences in pilot physical and mental workload imposed by environmental factors in the simulator when compared with those in the aircraft are important for several reasons. First, they give a comparative picture of environmental fidelity as it affects pilots in the simulator and in the aircraft. Flight tasks under comparable weather factors in the simulator and in the aircraft should result in comparable workload measurements, as should airspace operational factors, such as heavy air-ground communications workload. Second, variations in environmental factors may allow instructors to examine how these factors affect crew performance. A pilot may perform adequately under benign weather conditions, but poorly under the higher workload imposed by deteriorating weather conditions. Knowing the relationship between these environmental factors and workload levels provides a potentially valuable evaluation tool, but only if the environmental factors have comparable effects on workload in both simulator and aircraft. Finally, pilots are known to alter or re-prioritize tasks as a means of adapting to high workload conditions. Less attention may be paid to planning and preparation for future tasks due to pre-occupation with tasks at hand.

Environmental factors, such as changing weather conditions, may affect pilot workload capacity which, in turn, will affect pilot task performance.

Problem-solving and decision-making behaviors are more difficult to measure and therefore are more difficult to compare between the simulated and real aircraft environment. Pilot instructors and evaluators typically measure these pilot behaviors based on how well the pilot uses available information, weighs the importance of that information, exhausts alternatives and uses available resources effectively. For this reason, the simulation environment's ability to reproduce these behaviors in a ground-based device will likely be determined by how well the device supports the instructor or evaluator's needs in creating realistic and relevant tasks scenarios. If the instructor or evaluator feels confident that the simulator can provide all of the relevant aspects of an operational task scenario which allows them to accurately assess pilot problem-solving and decision-making skill, then the simulation environment will have successfully met this element of perceived fidelity. Many LOS simulation environments are now constructed in this way. The measure of the simulator design's ability to meet the criterion of perceived fidelity is based on how well it supports the creation of these complex LOS environments. In this area of perceived fidelity, perhaps more than any other, the simulator designer needs the assistance of experienced instructor and evaluation pilots to assist in defining the simulated task environment.

Fidelity and Training Transfer

Few beliefs are as strongly held or have generated as much controversy as the belief that physical fidelity of a flight simulator is directly correlated with its training value. If the simulator looks and functions like the aircraft it will be effective as a training device. Conversely, a device that does not look or function like the aircraft, it is believed, will *not* be effective in training a pilot. As we have seen previously, devices low in physical fidelity, such as the PCATD, have been shown to provide effective training. So there must be more to the issue of simulator fidelity and training transfer than the physical fidelity of a device.

Beyond the academic argument, however, there is a practical need to be able to determine the degree to which skills learned in devices of varying physical or perceived fidelity will transfer to the aircraft. One practical issue is the cost of the device. Cost is a major factor in training device development and deployment, and higher cost usually means that fewer devices will be available for pilot training. Answering the fidelity question would also address another practical issue important in civil aviation and that is the amount of loggable time a given device will be able to provide. The most objective and reliable answer to the question of training transfer and fidelity is found in transfer-of-training studies. However, these studies are not a practical solution given the required time and cost to conduct them. A better system of assessing fidelity and training transfer would incorporate a conceptual and analytic framework with which to determine the transfer effectiveness of a device.

Defining Training Transfer

To understand training transfer from one device such as a simulator to another device, such as an aircraft, requires some understanding of the basic mechanisms involved in the learning and remembering of skills and knowledge. It is well-known that human memory stores physical features of an object or event for only short periods (Tulving and Thompson, 1973). That is, the specific shape of an instrument or control, their specific location and other attributes are quickly forgotten. Since remembrance of training events requires long term storage of information, the lack of an ability to store physical details of events like a tape recording device would seem to be a liability. Such a recorder-style memory would, however, quickly overwhelm the brain's capacity to store information. Instead, human long-term memory is designed to integrate new information into an existing skill and knowledge store. Existing patterns of stored information are continually subject to alteration from incoming information if that information is in some way related to it. This integration process makes learning new skills easier if they have some relationship to older skills. It also makes retrieval errors for some information more likely. For example, it is not uncommon for pilots to revert to old skills appropriate to a previous aircraft type when transitioning to a new aircraft, particularly if the pilot is under great stress. In contrast, new skills and knowledge which have little or no relationship to those already in a pilot's long-term memory, will take longer to acquire, but will be less subject to retrieval problems when they are needed later.

Transfer of New Skills

In the primary phase of pilot training, training transfer is concerned with relatively new skills and knowledge. The training may take longer than more advanced pilot training, but is much less subject to interference from existing skills and knowledge, which may not be appropriate to the new environment. A device which supports this type of training only needs enough fidelity to allow for the formation of basic, generalizable skill sets. These include the relation between pitch attitude and airspeed, the effects of airspeed on control responsiveness, and other similar skills. The peculiarities or idiosyncrasies of particular aircraft characteristics are much less important in a primary device since the skills will easily generalize to a given aircraft type provided the basic relationships learned in the training device are still applicable. For the simulator designer, reproducing the exact spatial layout of instruments is much less important than maintaining the correct relationship between control inputs and instrument response, for example. So it should not be surprising that a PCATD with instruments much smaller in display size than the aircraft would still provide effective instrument training. Nor should it be surprising that controls (such as computer mice) whose response pattern bears little relationship to aircraft controls are much less effective or not effective at all.

Basic control skills for an aircraft, such as the ability to control airspeed, aircraft attitude, and so on, are not the only skills learned. Procedural skills, skills which require the execution of tasks in a specific order, are common flight tasks. While some procedural tasks are aided with the use of checklists or approach plates, many procedural tasks depend on pilot memory for their proper execution. For example, instrument approaches and holding pattern entries require a common procedure called the '5 T's' (Turn, Time, Twist, Throttle and Talk), each step in the procedure requiring a somewhat different interaction with the aircraft. However, the execution of the procedure is not dependent on the particulars of the aircraft. The procedural skill can be executed in almost any aircraft type. The simulator that supports the training of such procedural skills consequently does not need to have a high degree of physical fidelity with the specific target aircraft type in which the procedure will eventually be carried out. Indeed, many procedures of this sort have been trained in relatively simple mock-ups of an aircraft cockpit (Caro, 1988). For procedural skills, it is not the individual response that is trained, but the ordered execution of a set of responses in a pre-determined sequence. The responses, such as executing a coordinated turn, operating the navigation receiver, and so on, are behaviors that should be trained independently of the particular procedure and before procedure training begins.

Representational Skills

Once the pilot has mastered the basic control and procedural skills needed to operate the aircraft and navigate it to a desired destination, the development of more complex skills can take place. Representational skills are a set of cognitive skills which allow the pilot to develop mental representations of the operational environment. One such skill, *visualization*, is essential to instrument navigation and consists of the formation of a mental representation or mental picture of the aircraft's position using available navigational instruments, charts, and other information. Visualization is a skill needed to maintain situation awareness with regard to the position of the aircraft relative to its destination as well as to potential airspace hazards.

Representational skills are also exemplified by the pilot's ability to construct mental models of systems operations, including aircraft systems such as engine, hydraulic and electrical subsystems. These models help the pilot in diagnosing and resolving problems with systems operation. By forming mental models of the various components of a system and how they interact, the pilot is able to more readily understand the reason for system malfunctions and is able to develop remedies to resolve or at least to compensate for system failures.

Decision-Making Skills

Perhaps among the most important of all skills a pilot will learn is the ability to find and use information to aid in making decisions and to weigh the relevance and importance of that information in choosing among alternatives. Development of decision-making skills, such as those involved in weather avoidance, airport diversions, and others, requires the exposure to a variety of aircraft, airspace, and weather conditions that allow the exercise of such skills. Decision-making behavior also includes resource management skills that can facilitate the flow of relevant information from available resources, including other crew members or ground facilities.

Maintenance of Skills

Once the fundamental control, procedural, and cognitive skills are acquired, maintenance of proficiency in these skills is required of the pilot throughout his or her career. For pilots who fly frequently, maintenance of basic skills is not a problem. Each aircraft flight serves as a refresher course in the fundamentals of aircraft flight. However, many pilots do not fly frequently and some skills are not exercised regularly even by those who do. The flight simulator serves as the means of maintaining skills that would eventually be lost without this recurrent training.

Skills maintenance impacts the effectiveness of training devices and training transfer of these devices in several ways. First, different skills deteriorate at different rates. Loss of manual control skills occurs at a much lower rate than procedural skills, for example. As procedural skills are lost much faster than manual skills, most recurrent training schedules for simulator training are based on the anticipated deterioration rate for procedural skills. Devices that support procedural skill maintenance do not normally require a high degree of control fidelity since control skills are not the focus of training. This means that complex and sophisticated control loading systems or other technologies needed to support high fidelity handling characteristics are normally not needed if only procedural skills are being refreshed.

Much of commercial air carrier recurrent training is devoted to abnormal or emergency procedures training in the belief that, as abnormal and emergency events are rare, the skills necessary to correctly diagnose and recover from these events will deteriorate with disuse. There is the further assumption that recurrent training needs a very high fidelity flight simulator in order to be successful. For reasons stated above, this level of fidelity is largely unnecessary if only procedural skill maintenance is required. Procedural skills maintenance primarily requires enough perceived fidelity to allow the appropriate responses to be executed in the correct order. However, some procedural skills also incorporate the need for the pilot to make correct control skill inputs at a critical point in the procedure. In this case, simulator fidelity needs to be sufficiently close to that of the aircraft to allow the control skill to be evaluated and, if needed, relearned.

Maintenance of representational skills is much less dependent on a high degree of physical fidelity then either control or procedural skills. Most visualization skills can be maintained without a flight training device of any kind, since what is learned is a mental abstraction or construction derived from information. What is needed is a simple representation of the information sources (e.g., navigation instruments) with their associated setting and a system that provides feedback as to actual aircraft position. Similarly, maintenance of mental models of aircraft and other system components can be refreshed with system diagrams and mock-ups that can be made dynamic with the use of computer generated depiction of system operation. In both the case of visualization skills and mental models, there is little need for high physical fidelity in the training device.

Maintenance of decision-making skills is also not dependent on high physical fidelity of simulators. What is important is the pilot's ability to use effectively the available information resources and to weigh the information correctly to achieve the best decision possible. In many cases, these types of skills are best maintained in environments where large numbers of complex scenarios can be generated so as to thoroughly expose the pilot to a wide range of conditions. Well-designed computer gaming programs which recreate the important information resources and consequences of pilot decision-making may be far more effective than large, expensive simulators in maintaining these skills.

The above discussion is not meant to imply that devices with high physical fidelity, such as the typical airline FFS, cannot be used effectively for maintenance of these skills. It does, however, suggest that these devices may not be the most cost-effective or efficient training device alternative in maintaining many of the skills discussed above. However, a major advantage of very high fidelity type simulators like the FFS category is their ability to simultaneously and dynamically simulate a very complex operational environment. This makes them particularly valuable as an environment for the *evaluation* of piloting skills in an operationally realistic environment where issues such as multi-tasking and crew resource management abilities can be assessed.

Transition Training

Beyond primary skills training and instrument training, much of a professional pilot's career will be spent in *transitioning* to new aircraft types and categories. A professional civilian pilot will typically train on a basic primary trainer and then move on to a more complex, multi-engine aircraft. Some pilots will then transition to turboprop aircraft, to small jets and finally to large jet aircraft.

Transitioning to different aircraft types will consume much of the professional pilot's training time and much of that time is likely to be spent in flight simulators. Transition training, or 'differences training' as it is sometimes called, is designed to provide the pilot with the skills required of the new aircraft type and to modify old, existing piloting skills to conform to the requirements of the new aircraft. A flight simulator designed to support transition training needs to focus on the differences between the new and old aircraft. If the pilot is transitioning from an aircraft with markedly different handling characteristics, then the simulator needs to provide very high levels of perceived fidelity in the new aircraft handling qualities so these new characteristics are quickly perceived by the pilot. Because transition training often involves *unlearning* old behavior, the simulator cannot simply copy features from the old aircraft and place them in the new aircraft simulator. For example, in order to train pilot control skills in single-engine operations in multi-engine aircraft requires quite different performance, flight dynamic and handling characteristics than a single-engine aircraft. It also requires the supporting engine instrumentation array since the pilot must determine which engine has failed—a task not required in a single-engine aircraft.

For simulator fidelity and the transfer of training to the new aircraft, the most important issue is to assure sufficient fidelity between the new aircraft simulator and the new aircraft so as to preclude the transfer of old, inappropriate skills. This goal is achieved at the simulator design level by maximizing the physical and perceived fidelity in those areas where differences are most important to the development of new skills and unlearning of old ones. Many Cockpit Procedures Trainers (CPT) and other part-task training devices are well-suited for this purpose. Focusing on what is different between the two aircraft and on the elimination of inappropriate pilot behavior will result in more effective and efficient simulator training.

Measuring Training Transfer

In an earlier chapter, it was noted that the most important measure of the effectiveness of a flight simulator or flight training devices is the amount of aircraft time saved for the time the pilot spent in the simulator, as represented by the TER. This measure necessitates the conduct of transfer-of-training studies in order to determine the aircraft savings. Such studies are very difficult and expensive to conduct and, as a consequence, the TER measure is of limited usefulness as a general measure of simulator effectiveness. Yet the need to conduct some sort of assessment of simulator training effectiveness will continue to persist, particularly as new training technologies and methods are introduced.

Two other methods are available. One method is particularly useful in overcoming in inherent difficulty of controlling operational variables in transfer studies. However, it does not measure training transfer as such, but transfer of a proficient skill from an aircraft to the simulator. Thus, *reverse* transfer studies will require pilot flight proficiency in the aircraft that is simulated. This same skill is then measured in the simulator. The degree to which the same pilot behavior or skill can be reproduced in the flight simulator is then taken as a measure of simulator fidelity (see Steurs, Mulder, and Van Passen, 2004). As reverse transfer studies measure the degree to which the skill of an experienced pilot transfers to the flight simulator, it makes such studies viable candidates for assessing the perceived fidelity of the simulator design. Their singular shortcoming, however, is that they do not measure the effectiveness of the simulator's instructional technology or its instructional design components.

Other methods, such as the quasi-transfer study, can capture more of the training value of flight simulators than the reverse transfer evaluation while avoiding some of the shortcomings of traditional training transfer studies. Quasi-transfer studies measure the degree to which pilot behaviors learned in low fidelity devices, such as PCATDs, transfer to higher fidelity devices such as an FFS. The quasi-transfer study is based on the assumption that the higher fidelity training device or simulator is an adequate aircraft surrogate. If training transfers from the lower fidelity to the higher fidelity device, it is believed that training in the lower fidelity device will also transfer to the operational aircraft. The quasi-transfer study is often suggested as an alternative to the transfer-of-training study because it does not involve the need to use actual aircraft. There remain, however, questions concerning the assumption that one can directly equate quasi-transfer and conventional transfer-of-training study results (Taylor, Lintern, and Koonce, 1993).

More efficient and reliable analytical methods are still needed to determine the training transfer one can expect from a given simulator configuration and training regimen. The reliance solely on empirical evaluations of transfer effectiveness, with the problems of cost and timeliness that they inevitably bring to the design process, means that very few studies are likely to be conducted. Reverse transfer studies, while not directly measuring the instructional component of a simulator, do allow a direct and reliable assessment of the perceived fidelity of a device. Such studies still require the use of aircraft, the operation and instrumentation of which is costly and time-consuming. Methods that employ reliable engineering models that are sensitive to the unique issues involved in human-systems integration are needed. Such models would allow simulation designers to assess and evaluate the cost and benefit tradeoffs that are made in the flight simulator design process.

Summary

The ability to determine whether the design of a flight simulator will provide for effective pilot training is essential. Simulator design goals which stress physical reproduction or physical fidelity provide a poor foundation for the inevitable design tradeoffs that occur during the development of flight simulators. The flight simulator designer's goal must be focused on creating an environment where pilots will behave in manner comparable to their behavior in the aircraft and whose experience in the simulated aircraft will be indistinguishable from that of the aircraft. This *perceived* fidelity between the simulator and aircraft increases the likelihood that the simulator will be an effective training tool because it concentrates the design effort on those aspects of the simulation that are intended to make the experience of the pilot in the simulator comparable to the aircraft rather than on the physical or functional reproduction of as much of the aircraft as cost and technology will permit.

Chapter 8

Limitations in Flight Simulator Design and Use

Introduction

Developments in flight simulation technology in the past few decades have been dramatic indeed. Anyone who has experienced the latest advancements in flight simulation can only marvel at the ingenuity of the engineers who design and build them. Yet there are aspects of flight simulation which persist as significant barriers to simulator development and use. For the most part, these are obstacles that are likely to remain as such for the foreseeable future. Understanding their fundamental characteristics will aid those involved in the development and use of flight simulators to better apply the simulation technologies available.

Simulator Sickness

One of the more pervasive problems encountered in the flight simulator's long development has not been technical, but human. Simulator sickness is a form of motion sickness which has its origins in the early development and use of wide FOV visual display systems. Also called 'Simulator Adaptation Syndrome' or SAS, its symptoms are similar to other forms of motion sickness. They include fatigue, sweating, dizziness, nausea, and vomiting. Some of the earliest evidence of significant occurrences of simulator sickness was reported by Havron and Butler (1957). They found widespread simulator sickness in pilots undergoing training in a wide FOV, US Navy helicopter simulator. As visual display technology advanced in the years that followed, more reports of simulator sickness emerged not only in helicopters, but in fixed wing aircraft as well. The symptoms of simulator sickness are particularly debilitating and have a significant impact on the utility of simulators that produce it. This is especially true for military pilot training where simulator sickness has had a significant impact on simulator use. Due to its importance to pilot training, a review of the causes, consequences, and potential remedies for simulator sickness are included here.

Simulator Design

The actual physiological or neurophysiologic reasons for simulator sickness are, as yet, unclear and no single theory is accepted for its occurrence. However, there are certain design variables that affect the likelihood of simulator sickness. One such design variable, a wide FOV_H, and a related factor, image detail in the visual periphery, may combine to generate significant optical flow patterns in the pilot's visual periphery. This most often occurs during low level flight maneuvers where visual scene details are the most complex over the largest area of the visual display. These same design factors are recognized as the sources contributing to the illusion of self-motion or vection in pilots. Indeed, there appears to be a strong connection between vection and simulator sickness onset (Hettinger, 2002).

The provision of wide FOV_H, highly detailed image displays in many flight simulators would seem to make them all candidates for inducing simulator sickness in pilots. However, the onset of this reaction appears to be limited to those devices which require significant and sustained maneuvering tasks under conditions where the simulator visual system is likely to produce significant optical flow patterns, particularly in the pilot's visual periphery. Significant optical flow field activity in the visual periphery is typical of flight tasks in helicopters and fighter aircraft, particularly those tasks which require low level flight. The low level flight tasks in wide FOV simulators where significant aircraft maneuvering takes place are likely to produce the most severe simulator sickness symptoms. The absence of sustained and significant maneuvering tasks which would expose the pilot to these precursor conditions probably accounts for the lack of reports on simulator sickness in more advanced commercial airline simulators despite their use of wide FOV visual systems. Commercial airline maneuvers do not normally include sustained or significant maneuvering tasks at low altitudes where exposure to optical flow fields in the pilot's visual periphery would be occur as a result.

One of the theories attempting to account for motion sickness suggests that a conflict exists between sensory inputs, and this conflict results in the physiological symptoms we call motion sickness. Following from this theory, it would seem possible that the syndrome might be due to the lack of associated whole body motion cues that would normally accompany changes in the visual scene optical flow pattern. Reviews of simulator sickness events in flight simulators, however, reveal that they occur in both fixed-based simulators and in those equipped with motion platforms (Kennedy, Lilienthal, Baerbaum, Baltzely, and McCauley, 1989). It is, of course, possible that the platform motion systems used did not provide adequate motion stimulation and therefore are not a viable test of the theory. However, the data do suggest that typical motion platform systems used in flight training devices are not the solution to the simulator sickness problem.

Exposure Time

While it is possible to reduce the severity of simulator sickness by reducing the FOV_H of the visual display, such an expedient will likely impact the utility of the simulator. Reducing the FOV_H requires a careful analysis of the tradeoff between the training or research value of the simulator and the likelihood of simulator sickness in the pilots using the device. Alternatives to changes in design have been examined. Reductions in the severity of symptoms have been found when exposures to the simulator environment allowed the pilots to adapt. Intervals of a minimum of two days between sessions have been found to be effective in reducing symptoms (Kennedy, Lane, Baerbaum, and Lilienthal, 1993). Minimizing exposure to significant visual field movement can also reduce the severity of symptoms. For example, interspersing periods of low level flight activity with tasks at higher altitudes or tasks in instrument conditions may help pilots adapt to the simulator and to the precursors of simulator sickness. This might seem an unwarranted interference but the effects of simulator sickness on the pilot's ability to learn and perform are often so significant as to warrant changes in training syllabi or research protocols in order to minimize the influence of the syndrome.

The effects of simulator sickness are not limited to the simulator session itself. Perhaps the most pernicious aspect of the simulator sickness syndrome are the aftereffects of the phenomena. Aftereffects, such as sensations of turning and problems in postural stability lasting from 8 to 10 hrs are not uncommon in some flight simulators (Kellog, Castore, and Coward, 1984). A more recent survey found simulator aftereffects lasting as long as 24 hrs in some pilots (Ungs, 1989). These aftereffects have obvious safety implications, particularly for pilots operating vehicles, including aircraft, after simulator sessions. Precautions need to be taken to assure that those who develop simulator sickness syndrome are made aware of the potential aftereffects of the phenomenon.

Simulator sickness symptoms can be reduced in their intensity but are likely to remain a serious problem in the use of some flight simulators for the foreseeable future. Developers and users of flight simulators with wide FOV_H visual scene simulation systems in task environments which require significant and sustained low level maneuvers need to examine whether such wide FOV_H is necessary for training and whether the training value offsets the potential problems created by the simulator sickness syndrome.

Motion Cueing

The issue of whole body motion cueing in flight simulators was dealt with in an earlier chapter. But the essential difficulty in ground-based motion simulation presents a serious obstacle to flight simulation fidelity if the goal is to replicate all aspects of aircraft flight. To apply the necessary accelerations to ground-based flight simulators in an attempt to reproduce all of the forces a given aircraft is likely to encounter in operational flight is essentially impractical and, for training devices at least, cost-prohibitive. There are examples, of course, where research devices have been designed to reproduce some portions of the translational or rotational accelerations of operational flight and have had limited success in doing so, but the practical difficulties of simulating accelerations in all axes of flight at a level that would match real-life accelerations possible in a particular aircraft remain. For flight training devices in particular, the design constraints placed on simulator motion cueing systems have been so great as to raise serious questions regarding their value in simulator flight training.

The most desirable path to improving the cost-effectiveness of motion systems is to examine the value of motion cues in the training and evaluation of pilots. This necessitates a detailed examination of the role specific maneuvering and disturbance motion cues play in the performance of flying tasks. Once this is known, the motion cueing device can be tailored to provide only that motion information required. In civil aviation, specific regulatory requirements dictate the use of 3-DOF and 6-DOF motion systems for certain types of training remain as an obstacle to this approach. For other flight training devices, however, this approach may be viable provided other elements of the device which are governed by regulatory requirements are not adversely affected by the addition of motion cueing systems. This method of defining motion cueing requirements is not only more cost-effective, but will allow the development and application of new motion cueing technologies other than conventional motion platform systems in use today.

For those demanding the full range of possible motion cues including those which approach the structural limits of an aircraft, ground-based devices are probably not the answer. The cost associated with true full fidelity motion cueing systems is rarely justified. An airborne test or training system may, in the end, be the only viable answer to requirements for full physical motion fidelity.

Adaptation and Compensatory Skills

One way of viewing perceived fidelity in a flight simulator is that a simulator with perfect fidelity does not require the pilot to adapt his or her flying skills to fit the design limitations of the device. Thus, an experienced pilot of a Boeing B-737-700 should be able to operate an aircraft-specific, B737-700 flight simulator without altering existing skills or adding new ones. Likewise, those trained to proficiency levels in the device ought subsequently to be able to fly the aircraft without the need to modify or unlearn skills acquired in the simulator.

Despite the advances of flight simulator technology, it would be difficult to find a device that so closely reproduces the perceived fidelity of a given aircraft that no adaptation by the pilot is required. Nonetheless, the goal of perfect perceived fidelity ought to be a requirement of any flight simulator design that wishes to maximize training transfer. With such a goal comes the question as to how to measure adaptation. One of the means to do is to identify the presence of compensatory skills.

In an early study examining the viability of simple cockpit mockups for training procedural skills, Caro (1988) found that the mockup, which had no functional controls, could be just as effective as a fully functional procedures trainer. The cost implications of such a finding are obvious. However, Caro noted that a variety of what he described as 'mediating' skills needed to be learned by the pilot trainee in order to complete the procedures training in the mockup. These compensatory skills then had to be unlearned when the pilot operated the aircraft. Caro reports that this was accomplished rather quickly and without apparent difficulty. However, the learning and unlearning of compensatory skills takes time and effort which could be devoted to more useful and relevant skills. Moreover, some training programs, particularly those based on zero aircraft training time, may not be able to risk the transfer of compensatory skills to the operational aircraft even if such skills appear to be relatively benign with regard to aircraft safety.

The need for a pilot trainee to develop a compensatory skill is an obvious indicator that the training device may be improperly designed. However, many forms of adaptation to a simulator's lack of perceived fidelity are less obvious and have unanticipated consequences for training transfer. As described in an earlier chapter, distortions to normal depth and distance perception can be easily introduced by visual scene simulation systems which do not take into consideration pilot perceptual processes. Significant process and control delays, particularly those that involve closed-loop control tasks, may also result in the development of compensatory skills.

Adaptation to the simulator anomalies can be hastened by various methods (see Welch, 2002) but the cost of these design shortcomings on the training value of the simulator needs to be weighed against whatever benefits the low fidelity device has to offer. First, adaptation, whether in the form of sensory or perceptual motor adaptation or in the form of compensatory skills, takes time and effort that the pilot could be using for other purposes. Secondly, the consequences of adaptations may not be readily observed in the simulator. They may only arise in the actual aircraft and under conditions that may prove hazardous. For example, significant simulator process and control delays between pilot control inputs and the response of the visual scene or simulated instrument can result in control behaviors wholly inappropriate for the aircraft simulated. Improper control behavior during critical flight phases, such as takeoff and landing, acquired in a poorly designed flight simulator is both hazardous and difficult to correct. Compensatory skills that transfer to the aircraft from the simulator can, of course, be unlearned with time and with help from an instructor pilot. This inevitably lowers the overall effectiveness of the flight simulator as a training device.

In general, requiring pilots to adapt to the peculiarities of a flight simulator should be considered as evidence that the simulator design lacks essential perceptual fidelity. This is particularly true if the pilots are already proficient in operation of the simulated aircraft. Avoiding the need for pilots to adapt to a distorted sensory world or to develop compensatory skills to overcome simulator design inadequacies should be considered an integral and essential element of the simulator design process. It should also be a part of any regulatory requirement for flight simulator approval.

User Motivation

Despite the extraordinary advances in simulation technologies, flight simulators are, after all, ground-based devices which are not intended to fly. The motivation of users, both trainee pilots and instructors, are affected by the knowledge that the simulator is an artificial environment fixed firmly to the ground. Despite every attempt to create an environment that will 'immerse' the pilot in the flying task, the simulator cannot expose the pilot to the inherent hazards associated with actual flight. Therefore, the training and evaluation of tasks whose performance may be dependent upon the real possibility of injury or death are necessarily compromised. For example, an aircrew evaluation scenario may include deteriorating weather conditions at the destination airport. In order to make the correct decision, the aircrew must make an assessment of the risks associated with continuing to the destination. Normally, such an assessment includes the safety issues as well as factors such as passenger inconvenience. In the flight simulator, there are no safety issues as such, so the pilot will tend to apply company guidelines and other non-safety issues more rigidly. This is partly because of the absence of real hazards and in part because the aircrew knows that their performance is being observed. There are, unfortunately, recorded incidents in which an aircrew will attempt a potentially hazardous landing in real life operations even though they would never do so in the simulator simply because they know they are being observed and evaluated. Because of these unrealistic motivational conditions of a pilot, flight simulators are unlikely to be useful in assessing risk behavior as it would occur in the aircraft. This does not mean that the processes and strategies involved in decision-making can not be evaluated in a simulator, only that the limits of the evaluations should be understood.

User motivations also affect the training value and utility of flight simulators. This is particularly true for instructors who use simulators for training and evaluation. Instructors who view the simulator as less than adequate often convey their concerns directly or indirectly to pilot trainees. Poor user acceptance of the value of a flight simulator can significantly affect the motivation of instructors to teach, and of pilots to learn, the flying tasks for which the simulator was designed. This user acceptance problem permeates many issues in flight simulator design that often are considered as simply engineering problems. The user preference for simulator motion platforms appears to be one such area.

The motivations of those being evaluated in a device also affect the way the simulator design is assessed. Individuals who are relatively inexperienced in the aircraft for which the simulator is designed are less likely to assign fault to the simulator's design (whether or not such a fault actually exists). Experienced pilots, particularly those undergoing evaluation in the simulator, will be much quicker to assign fault to the simulator for their own performance problems.

Certain expectations may also affect pilot behavior. Because simulators are used extensively in pilot and aircrew performance and proficiency evaluations under other than normal operating conditions, pilots have come to expect failures in flight simulator training sessions that would not be given a second thought in the operational aircraft. Engine failures in jet aircraft are very rare, for example, but simulator engine failure scenarios are routine in training. Pilots can become highly motivated and attuned to failures in the simulated aircraft and this fact itself makes their performance perhaps significantly better in the simulator than it might otherwise be under similar circumstances in the aircraft. Those responsible for pilot evaluations in simulators should adjust the performance criteria for pilot success to reflect this fact.

System Architecture

Flight simulators and similar flight training devices are very complex devices that are essentially handmade from largely custom components. The systems architecture of these devices is generally closed to outside equipment designers. Component developers do not have available an open architecture which will support advances in technologies the way, for example, the PC industry has evolved. The comparison of simulator technology with the personal computer technology industry shows the glaring shortcomings of closed system architecture. Many areas of simulation technology such as visual scene simulation and communications lag far behind developments in the PC industry. While modern simulators are now taking advantage of the lower cost PC processing capability, much of the technology used in simulators is still proprietary. This is true of hardware and software components alike. The consequence of this closed systems architecture is not only high costs, but also the inability to take advantage of rapid changes in technologies and in software development. Advances in graphics, technologies, artificial intelligence, speech technologies, and other software areas have made many existing flight simulator designs obsolete. Moreover, users who have invested large sums of training capital in flight simulators cannot take advantage of commercial-off-the-shelf (COTS) components as they cannot be integrated within existing proprietary architectures. As simulators are often in use for ten years or more, the ability to rapidly and inexpensively upgrade existing simulators is essential to meet changes in aircraft avionics and other technologies.

There are a variety of economic and other reasons for the closed system architecture of the flight simulator industry, but the model assures a dwindling supply of increasingly costly devices to the aviation community. An open system architecture with appropriate industry standards would allow for lower development costs and ultimately much greater access to high fidelity devices by the training community. More importantly, access to COTS technologies would allow more rapid improvements in simulator fidelity, training effectiveness, and overall value to the user community.

Cost

While technical and user factors affect the utility of flight simulators, their high cost still remains a major obstacle to their widespread use in the aviation community. Not only the purchase cost, but also the cost of facilities, maintenance and necessary upgrades can place significant economic burdens on training and research organizations that are ill-equipped to manage them. Since pilot training is a cost and does not directly contribute to the profit of a company or to the completion of a military mission, it is subject to continuous cost-cutting efforts. Flight simulators and related training devices must be able to justify their costs by their contribution to the mission of the organization, whether civilian or military. The most direct contribution a training device has to such organizations is to reduce training time in the operational aircraft. The second direct benefit is to reduce the rate of accidents and incidents and to improve operational efficiency.

For commercial airline and military organizations, the benefits of flight simulator training far outweigh the costs associated with these devices. This is because operational aircraft ownership and operation costs are so high for these organizations that flight simulator use in place of aircraft use is readily justified. For example, at the time this book was written, a commercial airline FFS costs approximately one tenth of the aircraft it replaces. Despite the low ratio of flight simulator to aircraft cost, these devices can easily exceed ten million US dollars. Nonetheless, the training management of these companies is mindful of the need to reduce costs whenever possible. For this reason, simulator manufacturers must continually strive to reduce manufacturing and development cost.

Other parts of the aviation community are even more hard-pressed to justify the high cost of modern flight simulators. In contrast to commercial and military training operations, many smaller charter and general aviation operations have only limited access to simulation technology. For these operators, the cost of the training device is more difficult to justify since the costs of the aircraft they operate are much less than those of commercial air carriers or the military. For example, the general aviation FTD is typically about one-half the cost of the aircraft it replaces. The operating costs of smaller propeller-driven aircraft are also much less than large, turbo-jet aircraft, so defraying these costs with ground-based simulator training is less important.

Some reductions in costs of FFSs and FTDs have occurred because of the advent of lower cost microprocessor technology. Many newer flight simulators depend upon PC processors for their operation. Additionally, motion platform systems which relied on costly hydraulically operated drive systems can now use much less expensive, electrically driven systems instead. Despite lower prices, only around a thousand FFS devices are in operation today in the entire world. While the exact number of FTDs in operation is not known, it is unlikely to be more than a few thousand. Given that there are hundreds of thousands of licensed pilots in the world, the likelihood of access by a pilot to this valuable training technology is relatively low. In general, only a very small proportion of pilots will ever receive training on the more costly FFS systems.

Regulation

Civil aviation is perhaps the most regulated activity or enterprise in existence. Government regulations and standards are applied at every level of civil aviation from aircraft air worthiness, airspace control, to pilot training and evaluation. It is not surprising then that flight simulator certification and use is also heavily regulated. The three current levels of flight simulator design, FFS, FTD, PCATD and the more recent Aviation Training Device (ATD) categories are all subject to government scrutiny in the US, Europe and elsewhere.

Government regulations are imposed on flight simulators at both the design and operations levels. At the design level, regulators apply standards of fidelity (usually *physical* fidelity) appropriate to the category of the device. Devices used primarily for instrument training, such as ATDs or FTDs, have to meet certain levels of physical fidelity needed for instrument flight. For example, basic instruments such as the attitude indicator and the directional gyro need to be provided and need to be fully functional. More advanced flight simulators, such as those in the FFS category, need to meet more stringent physical fidelity requirements such as realistic visual scene simulation, motion cueing, and ground handling characteristics.

Regulations are also applied to flight simulators when in operation. The main regulatory impact on simulator operations is the rule restricting the 'loggable' time that a given device is authorized. Loggable time is the time spent in the simulator that is actually loggable or creditable for a particular rating, license, or currency requirement. For example, FTDs typically allow at least 20 hrs of simulator time as creditable to the total 40 hrs of instrument flying time required for the instrument rating. Regulations, however, require that the time is loggable only if a certified flight instructor is present to instruct the pilot while in the simulator. The requirement that a certified flight instructor or check airman be present while the simulator is used is a common rule for loggable simulator time. Regulations are also applied to the recurrent maintenance of flight simulators. These regulations are intended to assure that device is being maintained within the original design specifications under which it was certified. In general, recurrent inspections of certified simulators are more rigorous for the larger FFS than for other categories. This is partly due to the use of FFS devices as zero time training systems, i.e., training systems within which the pilot will receive *all* of his or her training and will receive no training in the operational aircraft.

Regulatory requirements for simulator design assure that minimum standards are met for those devices that are intended to replace some or all of the aircraft time normally required to meet certain pilot training standards. However, regulatory requirements for simulator design tend to slow technological progress considerably. This is due, in part, to the slow process of simulator certification by the regulatory agency involved. This slow pace of certification makes the introduction of new component technologies and design concepts a difficult and lengthy process. Additionally, regulatory requirements for the certification of simulators adds significant risk to the development process since the simulator may not meet the certification requirements or may do so only with significantly added cost. The added risk of certification delays and costs makes long term investment in simulation design and manufacturer more difficult to justify.

Legacy Devices

Changes in regulatory standards for simulators must contend with the large number of existing or legacy devices, some more than two decades old, that exist in the operational aviation community. Significant changes in standards which might advance the effectiveness of flight simulators are necessarily difficult to implement as they risk de-certifying large numbers of devices already in use. The difficulty of upgrading or replacing old training devices as a response to changes in regulations is significant for training organizations, and regulators are aware of these difficulties. The natural tendencies of regulators then is to proceed very slowly with design improvements to flight simulator technology rather than moving forward as soon as new technologies or designs become available. The consequence of this approach to simulator design is that current technologies used in simulators often lag far behind technological advances in many areas.

Modifying regulatory requirements in any substantial way is inherently difficult. However, the process of certifying flight simulators needs to be improved if it is to avoid becoming an obstacle to cost-effective system development. One way of doing so is to shift regulatory requirements away from often unsubstantiated design specifications to a performance-based, design evaluation. In this approach, a set of design criteria of a flight simulator are based, not on what is technologically available at a given time, but what has been demonstrated to be important to specific pilot training and evaluation syllabi to produce competent pilots for a particular type of operation and aircraft category. If it is demonstrated, that, for example, motion cueing is not important to the training of airline pilot in center-thrust turbojet aircraft, than those operators engaged in this type of training should not be required to purchase simulators with motion cueing capability. A performance-based, design criterion could be applied to other simulator components as well, including dynamic control loading, wide FOV visual scene simulation, visual scene display effective resolution, and others. By tailoring simulator design criteria to a demonstrated need, flight simulation technology developments are more likely to produce cost-effective solutions.

Summary

Despite the advances in flight simulator technology in the past decades, the design and use of simulators have inherent limitations. The effects on the utility of flight simulators, such as simulator sickness and user motivation as well as the limits on motion-cueing and the development of compensatory skills, are all problems of human-systems integration, which need to be considered when designing and using these devices. Other limits on flight simulator technology, such as cost and regulatory factors, have broader implications for the advancement and availability of this technology. The latter represent significant impediments to the use of flight simulators in the aviation community.

Chapter 9

Advances in Flight Simulation

Introduction

Throughout the history of flight simulator technology there have been periods of revolutionary change followed by periods of incremental improvements. The last period of revolutionary change came with the introduction of microprocessor technology and the rapid growth in computational power it made available. Much of flight simulation technical developments since then have been incremental in nature with small, but significant enhancements in visual display, motion platform, sound simulation and other areas. Some advancements in flight simulation are more noteworthy than others because they serve as a foundation for major improvements in the value of flight simulation for training and research. In this chapter, some notable advances in technology are examined which are now being used in flight simulators or have the potential to be used in the future. These technologies generally were not developed for flight simulators, but for other, often unrelated, purposes.

Image Generation and Display

Since the earliest days of flight simulation, the desire to create dynamic, realistic visual scenes for flight simulator displays has been persistent. Among the earliest attempts in image generation before the advent of the microprocessor revolution, was use of model boards scanned with high resolution cameras (see Figure 9.1). The model board was scaled to a fixed size covering the specific terrain or gaming area of interest. Scaled 3-D models of objects such as airports, buildings, ground vehicles, and others were built to the same scale and placed on the terrain model board. A camera with a variable zoom lens was placed on a gantry capable of moving in the same three axes of motion as those of an aircraft. The images generated from the camera as it moved over the model board were then displayed on a monitor or projection system in front of the pilot. Such systems had definite shortcomings not the least of which was very limited variations in visual imagery and in the size of the gaming area that could be displayed. Such systems were, however, the only systems capable of providing the level of detail and flexibility needed until the later development of the modern CGI systems.

Figure 9.1 Early Camera Terrain Model for Scene Generation
(Image Courtesy of NASA)

Satellite Imagery

Subsequent CGI developments have dramatically expanded the size of gaming areas in visual scene simulation as has the development of advanced visual texturing systems. Nonetheless, very large gaming areas needed for VFR cross-country flight training or long distance military missions involving low level flight are difficult to accommodate even with CGI systems. It is still difficult and costly to build large-scale, highly detailed gaming areas with existing CGI technology.

An alternative to CGI and texture fill technologies has been the use of satellite imagery. Until recently, the high resolution imagery from earth orbiting satellites was restricted to government use. With the advent of commercial satellite imagery, however, high resolution imagery has now become available to the general public and to the flight simulator designer. Because satellite imagery is taken from orbiting satellites hundreds of miles in space, images covering hundreds of nautical square miles are possible. The level of detail on a satellite image is dependent on the orbiting altitude of the satellite and, most importantly, on the satellite camera resolution. Resolution of satellite camera imagery is measured in meters per pixel. A 5m/pixel resolution represents a capability of resolving an object of 5 m², a 1m/pixel resolution image can resolve an object of 1 m², and so on. Two samples of satellite imagery of different levels of resolution are shown in Figure 9.2.

(a)

(b)

**Figure 9.2   Satellite Imagery of Los Angeles International Airport at
(a) 4 Meter Resolution and (b) 0.25 Meter Resolution**
(Images Courtesy of US Geological Survey)

An important aspect of satellite or any other photo imagery used in simulations
of visual scenes is that the image resolution determines the minimum altitude at

which the scene will appear realistic. Since digital images have a fixed resolution, the simulator image will appear blurred at altitudes below this minimum. This blurring or pixilation of the image is the same effect produced by 'zooming' in on a digital image. Unlike CGI 3-D object images, photo images have no vertical dimension. Such images need to be combined with other imaging and terrain modeling techniques in order to achieve 3-D effects. The addition of elevation data to these images is particularly important for use in flight simulators in order that image details are rendered at elevation levels which correspond to published aeronautical charts.

With digital satellite imagery and the use of various terrain modeling techniques, strikingly realistic visual scene imagery can be provided in flight simulators. In Figure 9.3, digital satellite imagery combined with digital elevation modeling reveals the details of the El Misti volcano near the city of Arequipa, Peru (the original image is in color). The volcano rises 5,822 meters above the city below it.

Figure 9.3 Satellite Imagery Combined with Terrain Modeling
(Image Courtesy of NASA)

Area of Interest Display

Many flight simulator visual scene display systems require only a limited FOV to meet the training or research purposes of the device. The display requirements of these devices are readily met by the available technologies that were described in Chapter 1. Some flight simulator applications are, however, more demanding with regard to both FOV and the resolution required of display systems.

The combination of very large FOV and high effective resolution is typical of fighter aircraft and some combat helicopter simulators. Both types of aircraft have very large FOVs available to the pilot since both are used in operations that require the pilot to detect and classify small targets at long ranges, either in the air or on the ground. Providing visual displays which can cover these large FOVs at very high effective resolution is both difficult and very expensive. The high cost of such display systems severely limits their availability to the flight training community. This, in turn, results in restrictions in access to the high fidelity flight simulators needed to support air combat training.

A more cost-effective solution has been developed which relies on the fact that only a very small area of the visual display needs to provide high resolution and that is the area that the pilot is viewing at a particular time, the area of interest. As much of the pilot's visual acuity is at or near the center of the eye, there is no need to have very high resolution throughout the display systems if the high resolution area could be provided only where the pilot was actually looking. This fact led to the development of what are called gaze-contingent multi-resolution displays or area-of-interest (AOI) displays. These AOI display systems allow the visual display of very high resolution scene simulation in areas where the pilot is looking and lower resolution in all other areas of the display. Where the pilot is looking is determined by the system through head-slaved or eye-slaved positioning systems that tell the image generation and display system where the pilot is looking and, hence, where the highest resolution is needed.

Ideally, the pilot using an AOI display system will not be aware of the fact that only the areas he or she is viewing directly have very high effective resolution. The image will maximize resolution in the center of the image and display a lower resolution image in the pilot's visual periphery. Because the operation of these devices must operate without pilot awareness, the mechanisms which determine the area the pilot is currently viewing must be both highly reliable and accurate. Since the image provides a multi-resolution image approximating that of the pilot's own visual system, the image presented by the AOI display should be perceived by the pilot as having high resolution everywhere in the display.

The computational savings of AOI systems can be very substantial. A recent review of AOI systems research has revealed that such systems can have a major impact on simulator visual system design as well as cost (Reingold, Laschky, McConkie, and Stamp, 2003). The review found image rendering times were reduced by a factor of 4 to 5, 2 to 6 times fewer polygons were required and 35 times fewer pixels were required than if a constant high resolution image were used to generate the visual scene.

AOI display systems are a less-expensive alternative to more conventional display systems where the requirement is to display very high resolution images over a very large FOV. AOI displays systems do incur higher costs to maintain due to the complex head-slaved and eye-slaved components that are required to operate them. However, they are an answer to a very specific need in the aviation training community. Whether AOI display systems are applicable to a specific task requires an analysis of that task to determine whether AOI systems are the best solution given the available alternatives. For those flight tasks that require effective display resolutions matching those of human visual acuity over very large FOVs, AOI systems are likely to be the most effective choice for the foreseeable future.

Computer Generated Instrumentation

From the very beginning of flight simulation development, reproducing the analog instrumentation of aircraft cockpits has proved a challenge for designers. In order to provide the necessary physical and functional fidelity, simulated aircraft cockpit instruments, such as attitude, altitude, airspeed and others, were often the same as those used in the aircraft but modified to allow ground-based operation. This basic design is still in use today in most flight simulators. The use of actual flight equipment in simulators is expensive and the complex motors and drive systems inside the instruments create simulator maintenance and reliability problems. They also make modifications to the simulator cockpit more difficult.

In recent years, the power of computer-generated imagery has been harnessed to address the problem of instrument simulation. Rather than using individual analog instruments as is done in the aircraft, digital instrument simulation simply displays the entire instrument array on one or more computer monitors mounted into the instrument panel. Digital images of the instruments are then displayed on the monitors in the position corresponding to that of the aircraft. An instrument panel fascia containing the various instrument control knobs and switches is then overlaid on the computer monitors.

Computer generated instrumentation not only eliminates the reliability and maintenance costs associated with conventional analog instrumentation, but allows much more rapid reconfiguration of simulator cockpits to accommodate the instrumentation of different aircraft types. Moreover, the rapid changes occurring in aircraft digital avionics technology often meant the obsolescence of older flight simulators that could not be modified to accommodate the new avionics. With computer generated instrumentation, modern flight simulators can be easily modified to display new avionics, thereby improving both the value and longevity of the device.

Control Loading and Handling Qualities

In an earlier chapter, the role of control loading systems in the simulation of aircraft handling qualities was discussed in some detail. The future of control loading developments is likely to focus on reducing the cost and complexity of these systems. This will make such systems not only more affordable for advanced simulators, but will bring improvements in handling qualities to lower fidelity flight training devices as well.

The main developments in control loading are likely to shift away from complex and cumbersome hydraulic control systems to those incorporating some form of electronic torque motor force command and force feedback systems. These force feedback systems have made rapid advances in the past decade spurred on by demands in the PC gaming market. The addition of microprocessor-based digital control systems and improved software engineering makes these new force feedback systems suitable for many flight simulator control loading applications. It is expected that many flight training devices for which the expensive hydraulic control loading systems could not be justified will now be able to provide more realistic aircraft handling characteristics with the application of these low cost force feedback systems.

Motion Cueing

Platform motion cueing systems are undergoing a modest evolution with the replacement of hydraulically-driven actuator systems with electronic torque drive actuators. This shift parallels the technology changes in control loading systems because both developments are being driven by the need to reduce simulator costs. Eliminating hydraulically controlled systems will also reduce maintenance issues associated with this technology.

More dramatic changes in whole body motion cueing systems will depend on changes to existing regulatory requirements for flight simulator motion cueing. Currently, there is little incentive to change the existing Stewart platform architecture as long as this design meets regulatory requirements. It is expected that 3-DOF and 6-DOF motion platform systems will persist as de facto motion cueing standards unless and until there is stronger support for regulatory change.

The support for change in the existing requirement should be based on scientific evidence for the effectiveness of motion cueing architecture rather than simply on user acceptance issues. While the latter are important for the marketing and sale of devices, user acceptance should not be used to support the regulatory status quo or regulatory change. An analysis of the requirements of motion cueing should focus on the specific aircraft operational training requirements and tailor the motion cueing system engineering accordingly. It is quite possible that many aircraft operational training needs can be met with either very limited motion cueing or not at all. Tailoring the motion cueing device design requirements means that motion cueing technology could be employed with far greater cost-effectiveness then is possible under existing regulatory conditions.

It is anticipated that technologies other than motion platform systems will see further advancements and will become more readily available in flight simulators of all types. This is particularly true of devices that can provide limited disturbance cues, including vibrotactile cues. These devices may be employed in training devices, not as a response to regulatory requirements, but as a means of enhancing the overall perceived fidelity of a device by trainee pilots and instructors. This will only occur, however, if the cost of such devices is low and their reliability and maintainability is very high.

Instructional Technology

Since flight simulators are used primarily as training devices, they are essentially a form of instructional technology. However, surprisingly little attention or effort has been spent enhancing the instructional technology that might improve simulator training effectiveness when compared with efforts to enhance simulator fidelity. The strong bias toward the latter is probably a result of the view that the simulator will be most effective in training pilots if it accurately replicates the aircraft simulated. Once again, the attention of the designer and flight training community has been focused on the physical fidelity of simulators to the detriment of their primary mission as instructional devices.

While progress has not been nearly as great in the instructional elements of flight simulators as other areas, some progress has been made. Typically, modern flight simulators used for training will have an Instructor Operator Station (IOS). The IOS is the primary user interface between the instructor and the simulator. The IOSs of current simulators are now able to apply digital computer technology to many areas of the IOS that earlier analog systems could not do. Using the IOS, instructors can now control all elements of the simulator as well as the simulated aircraft. Simulator training scenarios can be preprogrammed before the training sessions to incorporate every aspect of a training scenario such as airports, weather, routes, simulated air traffic and other scenario components. The instructor can pre-program aircraft system failures or introduce them manually during the training sessions. Most IOS systems will also provide the instructor with the opportunity to monitor aircraft state parameters and other simulator data as well as record these data for later playback.

Computer Assisted Instruction

The modern IOS provides a potentially powerful tool for the flight instructor to manage the simulator training session. However, the training value of the flight simulator is still largely dependent on the skill of the instructor in the use of the technology. While the contribution of instructor skill to ultimate training effectiveness of flight simulators is not known, it is probably considerable. But the contribution is dependent on the teaching skill of the instructors and that skill can vary considerably.

The flight simulator or flight training device is a computer-driven, synthetic training system. As such, it is well suited to computer-based training technologies that could substantially improve the cost-effectiveness of flight simulators used in training. These technologies include computer-assisted instruction (CAI) which, in combination with developments in artificial intelligence, can be brought to the flight simulator training environment to assist in the training of many routine flight tasks. Intelligent CAI, or simply ICAI, could substitute for the instructor on many of the simpler, repetitive training tasks for which flight simulators are often used. Many instrument training tasks, such as holding patterns, radio navigation, instrument approaches, and others are candidates for ICAI systems. The essential criteria for the application of ICAI systems is 1) that task performance success can be objectively defined, 2) that performance can be reliably monitored and recorded by computing devices, 3) that meaningful trainee performance feedback can be provided by the system and 4) that the instructor can manage and supervise the trainee's progress on the task. The latter is particularly important for two reasons. First, regulatory requirements for logging simulator training time generally require instructor supervision at some level. Second, instructors are legally liable for the instruction a trainee pilot receives and need to be aware of trainee performance problems if any exist.

The third criterion, performance feedback, is essential to any learning environment. The trainee pilot needs to know if the task has been performed successfully, where errors were made, and how to best correct those errors. Candidate ICAI systems should contain enough knowledge to not only identify performance problems, but provide meaningful feedback to the trainee pilot on how to improve his or her performance. In conventional ICAI systems, such feedback is usually in the form of displayed text, but in flight simulators the use of advanced voice display systems may be more appropriate. Additionally, the use of advanced speech recognition systems may be employed in ICAI systems to allow trainees to make queries of the system using normal speech rather than relying on devices such as computer keyboards or touch screen devices.

Initially, ICAI systems in flight simulator training may find their greatest utility in recurrent or refresher training of tasks where a flight instructor is either not needed by regulatory requirement or because the training requirement is much less demanding. They may also be more suitable for procedures trainers or FTDs, where the cost of the instructor relative to that of the cost of the device is high and, therefore, where reducing the presence of an instructor could return higher proportional savings.

Visual Display Augmentation

For those flight simulators that provide visual scene simulation, the possibility exists of using the flexibility of CGI systems to augment the visual scene display to enhance the training of visual flight tasks. While visual simulations are primarily designed for realistic display of real world scenes, they have the capability to do much more.

An example of the use of visual display augmentation to improve pilot training is provided in a study by Lintern, Roscoe, Koonce, and Segal (1990). In this study, students beginning their basic flight training were given training in a visual flight simulator for the approach and landing task following sessions in a ground-based trainer that had no display system. These sessions were followed by additional training with a visual display attachment. For one group, the normal visual scene simulation display was augmented with a series of visual approach height guidance cues aligned to each side of the localizer course. This group also received a prediction cue in the form of an aircraft symbol velocity vector[9] overlaid on the visual scene. A second, control group also received training with a visual display, but with no visual cue glide path augmentation or velocity vector cue. The two hours of training with the augmented visual display system saved an average difference of 1.5 hrs of flight time (9.8 landings) in the aircraft when compared to the control group. In this study, visual display augmentation significantly enhanced the training effectiveness of the flight simulator for very little cost.

Visual display augmentation has potentially great value in improving simulator training effectiveness for those visual flight tasks where novice pilots are having difficulty identifying the appropriate visual cue for a given visual flight task. Experiments with augmentation cues in other visual flight tasks are needed to determine additional benefits of this technique for flight simulator training. Military flight simulation training of visual flight tasks might be enhanced significantly by the use of visual display augmentation especially for air combat training.

It may also be possible to use visual display augmentation to enhance flight simulators with low fidelity visual scene simulation. Rather than providing the necessary scene fidelity, which might incur considerable cost, augmentation of the scene with artificial cues could be used instead. However, the potential of augmentation cues in visual displays used for this purpose needs to be carefully examined for possible negative transfer effects.

Summary

Advances in flight simulator visual scene simulation are likely to include greater use of satellite imagery and terrain modeling in an effort to provide larger visual gaming areas with high levels of image detail. For those systems that require very large FOVs and effective resolutions which are limited only by the resolving power of human eye, some form of gaze-contingent or AOI multi-resolution technology will be used for the foreseeable future. Handling qualities will see improvements, particularly for lower cost FTDs and similar devices as force feedback systems using microprocessor and electronic torque motors become more prevalent. Increases in the effectiveness of training simulators with the use of intelligent CAI and other instructional technology, such as visual display augmentation, are also possible.

---

[9] A velocity vector symbol tells the pilot where the aircraft will intersect the ground or another point in space if no changes to the aircraft velocity vector are made.

Chapter 10

Flight Simulation in Research

Introduction

Flight simulators have been designed and developed for the past seventy years, primarily for the training and evaluation of pilots. In this regard, the flight simulator represents a highly sophisticated and complex technology which has been largely successful in its goal of creating a ground-based, synthetic environment for pilot training. But no discussion of flight simulation would be complete without at least some reference to its role in aviation research.

Unlike flight simulators used for training purposes, those used in research have demanded very high levels of physical fidelity. Generally, such simulators fall in two general categories. The first category represents those purpose-built to support a particular research program such as the study of aircraft handling qualities or the optimum design of a flight deck. A second category of research simulators consist of those that are modified versions of simulators which were originally design for training. These devices have the advantage of lower cost and the high reliability built into training devices. They are especially useful in research programs where extensive modifications or changes to the simulator design are not required.

Flight simulators used for research also have a different user population. Rather than the instructor pilots and trainees for whom training simulators are designed, the research simulator is typically designed and used by research scientists and engineers. The device itself is more likely to be flown by highly experienced pilots who will be more demanding of fidelity requirements than trainee pilots are likely to be. Additionally, research simulator hardware and software need to be easily changed to allow for comparative testing of either aircraft or simulator design characteristics. Easily altering the handling properties of a simulated aircraft or changing the update rate of the visual scene display are the types of changes that might be required of a research simulator. Research simulator design requirements differ significantly in this regard from training simulators where the design is fixed and changes may be not only be impractical, but prohibited by regulations.

Accurate and complete recordings of simulated aircraft performance under a wide variety of scenarios require extensive instrumentation of flight simulators used for research. This instrumentation may involve the recording of a large number of aircraft and simulator performance parameters at a very high sampling rate. Additional recordings of pilot or aircrew behavior are also required in some research studies requiring modifications to the simulator to accommodate audio, visual and other recording systems.

The designers of research simulators also need to address the issue of fidelity. Fidelity in research simulators is, however, even more important than it is in training simulators since much of the research conducted on these simulators is based on the assumption that the simulator is a wholly valid surrogate for the aircraft. While only partial transfer-of-training from a training device may be acceptable, the ability to generalize to the operational aircraft conclusions drawn from studies conducted in research simulators must be unquestioned for the investigator to have any faith in the utility of the research device. This requirement for very high fidelity in research simulators increases not only the required complexity and robustness of its component technologies, but significantly increases the cost of these devices as well. For this reason, purpose-built research simulators are relatively few in number and are generally owned and operated by government agencies or large corporations.

Research Applications

The role of the flight simulator in aviation research is highly varied, involving virtually every aspect of aircraft operations and every element of the aviation community. Every major civilian and military airframe manufacturer uses research simulators in the process of aircraft design and evaluation. Avionics developers also use simulators to test new concepts before installing them in operational aircraft. Research simulators are also used to evaluate handling characteristics of new and modified aircraft and they are used extensively to study crew and pilot factors in the design of cockpits and in efforts to improve pilot training. Research simulators also play a role in evaluating proposed changes to airport configurations including proposed improvements to airport runway designs. Military aviation uses flight simulators to evaluate weapons and tactics in simulated air combat missions. These and other research applications of flight simulators demonstrate not only the value of the technology, but the confidence of the research community that the technology faithfully represents a given aircraft and its operating environment. To illustrate in more detail the uses of flight simulation in aviation research, several research applications of the technology are reviewed in the following sections.

Cockpit Design

With the advent of powerful microprocessors and improvements in visual display technologies, both civil and military cockpits have undergone dramatic changes in the past decades. In civil aviation, commercial airline cockpits were designed for decades to accommodate a crew complement of two pilots and one flight engineer. With the new cockpit automation, which resulted from advances in microprocessor technology, the possibility arose of eliminating the need for the flight engineer altogether. Flight simulators were instrumental in aiding designers in the complex flight deck reconfiguration issues that allowed for flight operations with only two pilots. Cockpit layout alternatives, new avionics systems, and new flight management and diagnostic systems were safely and efficiently evaluated in ground-based simulators. Flight simulators were particularly valuable in allowing pilots the opportunity to 'fly' the new cockpits before the aircraft was built.

Military cockpit design was also affected by developments in computer and display technologies but the designs were also influenced by a new technology called 'fly-by-wire'. This technology replaced the conventional cabling and hydraulic systems for operating aircraft control surfaces, such as elevator and ailerons, with computer-controlled electronic actuators that were connected to pilot controls with simple, lightweight wiring. The pilot no longer operated the control surfaces directly but merely sent control signals to a flight control computer which did all that was needed. This design eliminated the need for cumbersome and heavy control devices in the cockpit and replaced them with much smaller control devices that could be placed in less space-consuming areas than conventional devices. Thus, the centrally-located cockpit joystick or yoke was eventually replaced by a side-arm controller. The cockpit designers could now take advantage of new display technologies and make substantial improvements to the overall design of the aircraft cockpit. Flight simulators played an important role in the development both of the side-arm, fly-by-wire controller technology and in the overall design of the military aircraft flight deck. Increasingly, cockpit designers now apply and evaluate new ergonomic design concepts for the flight deck by using flight simulators rather than an aircraft.

Avionics

The same advances that changed the design of modern aircraft cockpits also affected cockpit instrumentation and display systems. The conventional analog instrument displays, which were a fixture of cockpits for generations, were soon replaced by digital image displays. The new 'glass cockpit' designs have become commonplace in modern aircraft. A number of research programs using flight simulators were employed and are still employed in the development of this new form of cockpit instrumentation. An example of a research simulator used for this purpose is shown in Figure 10.1. Research programs now focus on improving individual display designs and on solving integration problems of these new displays with existing avionics suites. Programs using research simulators are also involved in the problem of information management that is associated with the continuous presentation of large amounts of information on the flight deck.

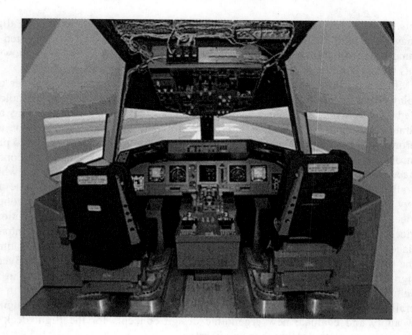

Figure 10.1 Flight Deck Research Simulator
(Image Courtesy of NASA)

New flight control display design concepts have also been aided by the use of research simulators. One of these is the Head-Up Display system or HUDs which is now commonly found on many military and some civilian aircraft. The HUD provides the pilot with critical flight control and other information displayed on a projection or transparent screen overlaying the forward cockpit window. The HUD thus allows the pilot to continue to view the visual scene outside the cockpit window while avoiding the necessity of looking down at cockpit instrumentation. For the civilian pilot, the need to transition from inside the cockpit to outside the cockpit for information during flight critical moments is thereby eliminated. In military aviation, the HUD also contains targeting and weapons system information, which the pilot can readily access during the high workload periods of air combat maneuvering. An example of a HUD is shown in Figure 10.2.

Research simulators are also involved in the development of new systems that will allow the pilot to see a synthetic version of the external visual scene on a flight display. This 'synthetic vision' uses a combination of technologies including infrared, radar, GPS and others to create and continually update a synthetic image of the external world. Such displays may prove invaluable under conditions where forward visibility is very limited or nonexistent. Research simulators are particularly valuable in the simulation of flight operations with these devices because of the inherent hazards of flying real aircraft in low visibility approach and landing operations.

Figure 10.2 Head-Up Display in NASA VMS
(Image Courtesy of NASA)

Airport Capacity Improvements

As commercial aviation continues to grow, the capability of existing airports to handle the increased air traffic has become a concern. One means of enhancing the capacity of an airport is to increase the utilization of its existing runways. Improved landing guidance systems like the Microwave Landing System (MLS) are a way to achieve this goal. The MLS approach guidance would allow some existing airports to use different approach plans and procedures thereby allowing access to runways that could not otherwise be used. Research simulators have been employed to examine how such approaches would affect the margin of safety when aircraft are flown by experienced pilots under operationally realistic conditions. Similar research programs have used simulators to evaluate the possibility of using parallel instrument approaches to runways that currently have less than the required separation for such approaches. Research simulators are also being employed in programs aimed at reducing runway incursions. Aircraft crossing or otherwise accessing active runways in use by other landing and departing aircraft have increased in frequency over the years as traffic density at airports has increased. Research simulators have proved valuable in evaluating new procedures and airport signage as well as advanced cockpit display systems designed to alert pilots of potential conflicts.

Human Factors

Increasingly, flight simulators are being used to address many safety issues that are a consequence of human factors. Human factors include issues such as pilot workload and fatigue, crew resource management, pilot skill maintenance, communications and poor human-systems integration. Human factors are a major cause of both commercial and general aviation accidents and are demanding increased attention from the research and regulatory community. Because the modern flight simulator is capable of providing a highly realistic flight environment, research into pilot and crew behavior is now possible. Pilot and crew behavior in complex and demanding environments require high levels of manual flying skill as well as problems-solving, decision-making and resource management skills. The results of these simulator studies will aid in developing improved training programs, operational procedures, crew scheduling and flight deck avionics.

Training Research

Improved training programs are, of course, one means by which these human factors problems can be solved. Flight simulators are a logical choice as a means of investigating training methods and technologies. Simulators have been used in this role for decades and continue to be used for this purpose today.

In recent years, extensive use has been made of simulators to investigate alternative methods by which crew coordination and communication training could be improved. The flight simulator has proved invaluable as a tool to examine problems of crew interaction. Studies using these simulators led to the development of modern crew training techniques such as CRM training. Modern CRM training programs are able to take advantage of the LOS training research also conducted in simulators more than twenty years ago (Ruffell-Smith, 1979).

Similar types of research programs have been carried out in military aviation to enhance the mission effectiveness of air combat training. The research simulator is being used, not only to improve CRM training methods, but also to examine improved methods of training tactical decision-making and problem-solving skills under simulated combat conditions. These conditions include ground-based and airborne threats as well as the requirement of the pilot to coordinate with other aircraft and ground-based control centers. The need for improved training methods in tactical decision-making represents an opportunity for flight simulation technology in both military and civil aviation training. Such research emphasizes the need for flight simulators to have the capability to provide both realistic aircraft simulation and the ability to simulate highly realistic operational task environments.

Accident Investigation

Flight simulators are also used as tools in safety research and accident investigations. The use of simulators has been aided by the installation of flight recorders on many aircraft. The recorders are capable of sampling hundreds of aircraft state parameters at very high rates. These data can then be used to re-create the aircraft state parameters in a flight simulator after the recorders are recovered from the accident site. In combination with other known data, such as weather and ATC communications, a real-time re-creation of the moments leading up to the accident are possible. The flight simulator allows the safety investigator to quite literally place him or herself in the pilot seat and experience what the pilot experienced just before the accident occurred. The ability to experience the accident event in this way makes this application of flight simulators particularly useful in investigating human factors issues in aircraft accidents. Use of the flight simulator in accident investigations allows the investigator to gain new insights into both systems and human causes of aircraft accidents, which could be obtained in no other way.

Flight Simulator Design

Not surprisingly, research simulators are also used in investigations into a wide variety of issues in flight simulator design. These include visual scene simulation, motion platform cueing, communications simulation, handling qualities simulation and other design issues. The research simulator is often used to determine which changes in one or more simulation component design parameters result in measurable changes in pilot performance. The rationale for such studies is that if differences in the design of simulator components have no measurable effect on pilot behavior in the simulation, then these differences are unlikely to affect pilot performance in the aircraft. This makes simulators a very cost-effective means of evaluating alternative designs for training devices by eliminating ineffective and potentially costly components of the training simulator design before the devices are produced.

Flight Simulator Research Facilities

Facilities engaged in aviation research where flight simulators are the principal research tool are located around the globe. Many facilities are owned and operated by government agencies and universities. A few facilities are operated by private industry, including airframe manufacturers and private research companies. The research programs in these facilities are widely varied in their goals. They include aircraft design and avionics research, aviation safety and security programs, human factors research, crew training research and research into improving aviation operations effectiveness. In the space available here, a few such facilities will be described. They have been selected to provide a sense of the wide diversity of research activities supported by flight simulator technology.

US Air Force Research Laboratory (USAFRL), Warfighter Training Research Division

The USAFRL Warfighter Training Research Division is located in Mesa, Arizona. Its primary mission is to conduct research into training methods and technologies in support of USAF missions. Research is conducted in a wide range of areas including crew training methods, training devices and simulation design. Research devices include multi-task training research simulators, unit level training research devices, and command, control and communications research simulators.

FAA Civil Aeromedical Institute

The FAA Civil Aeromedical Institute (CAMI) is located in Oklahoma City, Oklahoma. It is the primary research facility for conducting aeromedical research programs in the US. The human resources research division within CAMI operates research simulators such as the Advanced General Aviation Research Simulator (AGARS) which is used to investigate a variety of issues including human factors and systems integration problems unique to general aviation aircraft.

Cranfield University School of Engineering

Cranfield University is located in Cranfield, UK, and its School of Engineering has a long history of engineering research using flight simulators. Research at Cranfield is ongoing in avionics, flight control systems as well as human factors. Cranfield's facilities include a fixed-base B-747 simulator, a general aviation simulator used for part-task training research called the Aerosoft Flight Simulator (AFS), and a supporting air-vehicle modeling and simulation environment (AV-MASE).

Delft University of Technology

Delft University of Technology, located in Delft, The Netherlands, has been involved in simulator research for many years. Current research is in support of large transport handling qualities, flight deck design, and simulator fidelity research. Delft facilities include the SIMONA Research Simulator (SRS), a generic aircraft simulator.

The National Aerospace Laboratory

The National Aerospace Laboratory (NLR) has facilities located in Amsterdam and Noordoostpolder, The Netherlands. NLR has a highly diversified aerospace research program in civil and military aviation. Facilities at NLR include full-mission fighter and air transport aircraft simulators. NLR's research program includes avionics and aerodynamic research as well as other areas.

National Aeronautics and Space Administration

The National Aeronautics and Space Administration (NASA) is the primary civil aerospace research organization in the US. It is engaged in a wide variety of aerospace research programs. Primary flight simulator facilities are located at NASA-Langley Research Center in Langley, Virginia and NASA-Ames Research

Center in Moffett Field, California. NASA-Langley facilities include a wide variety of flight simulators including generic flight deck research simulators.

NASA-Ames Research Center facilities include the NASA VMS described in an earlier chapter and the Crew-Vehicle Systems Research Facility (CVSRF). The CVSRF was developed primarily to examine human factors issues in a variety of civil air transport domains. The facility includes a B-747 flight simulator and generic advanced flight deck research simulator. The CVSRF also incorporates an ATC simulation laboratory used to support flight simulation studies.

Summary

Flight simulators are used, not only as devices to train and evaluate pilots, but to support a wide range of aerospace research programs. These simulators include those which are purpose-built to support a particular line of research and others which are modified training simulators designed to support a broader range of research studies. Research simulators are often the first to employ advanced simulation technologies in support of investigations. These simulators are a vital part of aerospace research conducted at laboratories around the globe.

Bibliography

Atkinson, R.C. and Shiffrin, R.M. (1968). Human memory: A proposed system and its control processes. In K.W. Spence and J.T. Spence (Eds.), *The Psychology of Learning and Motivation* (Vol. 2). New York: Academic Press.

Bell, H.H. and Waag, W.L. (1998). Evaluating the effectiveness of flight simulators for training combat skills: A review. *International Journal of Aviation Psychology, 8*, 223-242.

Biggs, S.J. and Srinivasan, M.A. (2002). Haptic interfaces. In K.M. Stanney (Ed.), *Handbook of Virtual Environments (pp. 93-115)*. Mahwah, NJ: Lawrence Erlbaum Assoc.

Bradley, D.R. (1995). Desktop flight simulators: Simulation fidelity and pilot performance. *Behavior Research Methods, Instruments and Computers, 27*, 152-159.

Brandt, T., Dichgans, J. and Koenig, E. (1973). Differential effects of central versus peripheral vision on egocentric and exocentric motion perception. *Experimental Brain Research, 16*, 476-491.

Bray, R. (1973). A study of vertical motion requirements for landing simulation. *Human Factors, 3*, 561-568.

Brown, J.S., Knauft, E.B., and Rosenbaum, G. (1947). The accuracy of positioning reactions as a function of their direction and extent. *American Journal of Psychology, 61*, 167-182.

Burke-Cohen, J., Kendra, A., Kanki, B., and Lee, A.T. (2000). *Realistic radio communications in pilot simulator training.* (DOT Technical Report DOT-VNTSC-FAA-00-13). Washington, D.C.: Office of Aviation Research.

Burke-Cohen, J., Soja, N.N., and Longridge, T. (1998). Simulator platform motion-the need revisited. *International Journal of Aviation Psychology, 8*, 293-317.

Caro, P.W. (1988). Flight simulation and training. In E. L. Weiner and D.C. Nagel (Eds.), *Human Factors in Aviation.* New York: Academic Press.

Clark, B. and Stewart, J.D. (1968). Comparison of three methods to determine thresholds for perception of angular acceleration. *American Journal of Psychology, 81*, 207-216.

Cooper, G.E. and Harper, R.P, Jr. (1969). *The use of pilot rating in the evaluation of aircraft handling qualities.* NASA TN D-5153.

Comstock, J.R., Jones, L.C., and Pope, A.T. (2003). The effectiveness of various attitude indicator display sizes and extended horizon lines on attitude maintenance in a part-task simulation. *Proceedings of the Human Factors and Ergonomics Society.* Santa Monica, CA: The Human Factors and Ergonomics Society.

Dennis, K.A. and Harris, D. (1998). Computer-based simulation as an adjunct to ab initio flight training. *International Journal of Aviation Psychology, 8*, 261-276.

Dixon, K.W., Martin. E.L., Rojas, V.A., and Hubbard, D.C. (1990). *Field-of-view assessment of low-level flight and an airdrop in the C-130 Weapon System Trainer (WST)*. US AFHRL Tech. Rpt. Jan Tech Rpt 89-9 20.

Guedry, A.J. (1976). Man and motion cues. *Proceedings of the Third Flight Simulation Symposium*. London: Royal Aeronautical Society.

Hart, S.G. and Staveland, L.E. (1988). Development of NASA-TLX (Task Load Index): Results of empirical and theoretical research. In P.A. Hancock & N. Meshkati (Eds.), *Human Mental Workload* (pp. 139-1830). Amsterdam: North Holland.

Havron, M.D. and Butler, L.F. (1957). *Evaluation of training effectiveness of the 2-FH-2 helicopter flight training research tool* (Tech. Rep. No. NAVTRADEVCEN 20-OS-16, Contract 1925). Arlington, VA: U.S. Naval Training Device Center.

Hays, R.T. (1992). Flight simulator training effectiveness: A meta-analysis. *Military Psychology, 4*, 63-74.

Hettinger, L.J. (2002). Illusory self-motion in virtual environments. In K.M. Stanney (Ed.), *Handbook of Virtual Environments*. Mahwah, New Jersey: Lawrence Erlbaum Associates.

Hettinger, L.J., Todd, T.W., and Haas, M.W. (1996). Target detection performance in helmet-mounted and conventional dome displays. *International Journal of Aviation Psychology, 6*, 321-334.

Hood, D.C. and Finkelstein, M.A. (1986). Sensitivity to light. In K.R. Boff, L. Kaufman, and J.D. Thomas (Eds.), *Handbook of Perception and Human Performance: Sensory Processes and Perception*. New York: John Wiley and Sons.

Howard, I.P. (1986). The vestibular system. In K.R. Boff, L. Kauffman, and J.P. Thomas (Eds.), *Handbook of Perception and Human Performance: Sensory Processes and Perception*. New York: John Wiley and Sons.

Jennings, S., Craig, G., Reid, L., and Kruk, R. (2000). The effect of helicopter system time delay on helicopter control. *Proceedings of the IEA 2000/HFES 2000 Congress*. Santa Monica, CA: Human Factors and Ergonomics Society.

Jorna, P.G.A.M. (1993). Heart rate and workload variations in actual and simulated flight. *Ergonomics, 36*, 1043-1054.

Kellog, R.S., Castore, C.H., and Coward, R.E. (1984). Psychological effects of training in a full vision training simulator. In M.E. McCauley (Ed.), *Research Issues in Simulator Sickness: Proceedings of a Workshop* (pp. 2-6). Washington, D.C.: National Academy Press.

Kennedy, R.S., Lilienthal, M.G., Berbaum, K.S., Baltzely, D.R. and McCauley, M.E. (1989). Simulator sickness in U.S. Navy flight simulators. *Aviation, Space, and Environmental Medicine, 60*, 10-16.

Kirwan, B. and Ainsworth, L.K. (Eds.). (1992). *A Guide to Task Analysis*. London: Taylor and Francis.

Kleiss, J.A. and Hubbard, D.C. (1993). Effects of three types of flight simulator visual scene detail on detection of altitude changes. *Human Factors, 35,* 653-671.

Kruk, R. and Reagan, D. (1983). Visual test results compared with flying performance in telemetry tracked aircraft. *Aviation, Space, and Environmental Medicine, 54,* 906-911.

Lawson, B.D., Graeber, D.A., Mead, A.M., and Muth, E.R. (2002). Signs and symptoms of human syndromes associated with synthetic experiences. In K.M. Stanney (Ed.), *Handbook of Virtual Environments.* Mahwah, New Jersey: Lawrence Erlbaum Associates.

Lee, A.T. (2003). Air traffic control communications simulation and aircrew training. *Proceedings of the Royal Aeronautical Society: Simulation of the Environment.* London: The Royal Aeronautical Society.

Lee, A.T. and Bussolari, S. (1989). Flight simulator platform motion and air transport pilot training. *Aviation, Space, and Environmental Medicine, 60,* 136-140.

Lee, A.T. and Lidderdale, I.G. (1983). *Visual scene simulation requirements for C-5A/C-141B aerial refueling part task trainer* (AFHRL-TP-82-34). Williams AFB, AZ: Air Force Human Resources Laboratory.

Levison, W.H., Lancraft, R.E., and Junker, A.M. (1979). Effects of simulator delays on performance and learning in a roll-axis tracking task. *Proceedings of the Fifteenth Annual Conference on Manual Control.* Dayton, OH: Air Force Wright Aeronautical Laboratories.

Lintern, G. and Koonce, J.M. (1992). Visual augmentation and scene detail effects in flight training. *International Journal of Aviation Psychology. 2,* 281-301.

Lintern, G., Roscoe, S.N., Koonce, J.M., and Segal, L.D. (1990). Transfer of landing skills in beginning flight training. *Human Factors, 32,* 319-327.

Magnusson, S.M. (2002). Similarities and differences in psychophysiological reactions between simulated and real air-to-ground missions. *International Journal of Aviation Psychology, 12,* 49-61.

Martin, E.L. (1981). *Training effectiveness of platform motion: Review of motion research involving the advanced simulator for pilot training and the simulator for air-to-air combat.* US AFHRL Technical Report. Feb No 79-51 29.

Martin, E.L. and Cataneo, D.F. (1980). *Computer-generated image: Relative training effectiveness of day versus night visual scenes.* USAF HRL Tech. Rpt. Jul No. 79-56 31.

Morrow, D., Lee, A.T., and Rodvold, M. (1993). Analyzing problems in routine pilot-controller communications. *International Journal of Aviation Psychology, 3,* 285-302.

Mulder, M., Pleasant, J-M., van der Vaart, H., and van Wieringen, P. (2000). The effects of pictorial details on the timing of the landing flare: Results of a visual simulation experiment. *The International Journal of Aviation Psychology, 10,* 291-315.

Nahon, M., Ricard, R., and Gosselin, C.M. (2004). *A comparison of flight simulator motion-base architectures.* Dept. of Mechanical Engineering, University of Victoria, Victoria, B.C., Canada.

Nataupsky, M. (1979). *Platform motion contributions to simulator training effectiveness: Study III. Interaction of motion with fields-of-view.* USAFARL Tech. Rpt. Nov No. 79-25 28.

Orlansky, J. and String, J. (1977). *Cost-effectiveness of flight simulators for military training: Vol. 1. Use and effectiveness of flight simulators* (IDA Paper No. P-1275). Arlington, VA: Institute for Defense Analysis.

Ortiz, G.A. (1994). Effectiveness of PC-based flight simulation. *International Journal of Aviation Psychology, 4,* 285-291.

Padmos, P. and Milders, M.V. (1992). Criteria for simulator images: A literature review. *Human Factors, 34,* 727-748.

Pfeiffer, M.G., Horey, J.D., and Butrimas, S.K. (1991). Transfer of simulated instrument training to instrument and contact flight. *International Journal of Aviation Psychology, 1,* 291-229.

Pierce, B.J. and Geri, G.A. (1998). The implications of image collimation for flight simulator training. *Proceedings of the Human Factors and Ergonomics Society,* 1383-1387. Santa Monica, CA: The Human Factors and Ergonomics Society.

Regan, D.M., Kaufman, L., and Lincoln, J. (1986). Motion in depth and visual acceleration. In K.R. Boff, L. Kaufman, and J.P. Thomas (Eds.) *Handbook of Perception and Performance: Sensory Processes and Perception.* New York: Wiley and Sons.

Reingold, E.M., Laschky, L.C., McConkie, and Stamp, I.M. (2003). Gaze-contingent multi-resolution displays: An integrative review. *Human Factors, 45,* 307-328.

Rolf, J. and Staples, K.J. (1986). *Flight Simulation.* Cambridge, U.K.: Cambridge University Press.

Roscoe, S.N. (1980). *Aviation Psychology.* Ames, IA: Iowa State University Press.

Ruffell-Smith, H.P. (1979). *A simulator study of the interaction of pilot workload with errors, vigilance, and decisions* (NASA Technical Memorandum 78482). Moffett Field, CA: NASA-Ames Research Center.

Ryan, L.E. Scott, P.G., and Browning, R.F. (1978). *The effects of simulator landing practice and the contribution of motion simulation to P-3 pilot training.* TAEG Report, Sep No 63 39.

Sanders, M.S. and McCormick, E.J. (1993). *Human Factors in Engineering Design.* New York: McGraw-Hill.

Schraagen, J.M., Chipman, S.F. and Shalin, V.L. (2000). *Cognitive Task Analysis.* Mahwah, New Jersey: Lawrence Erlbaum Associates.

Sherrick, C.E. and Chulewiak, R.W. (1986). Cutaneous sensitivity. In K.R. Boff, L. Kaufman, and J.P. Thomas (Eds.), *Handbook of Human Perception and Human Performance: Sensory Processes and Perception.* New York: John Wiley and Sons.

Shilling, R.D. and Shinn-Cunningham, B. (2002). Virtual auditory displays. In K.M. Stanney (Ed.), *Handbook of Virtual Environments*. Mahwah, New Jersey: Lawrence Erlbaum Associates.

Stephens, D. (1979). Developments in ride quality criteria. *Noise Control Engineering, 12*, 6-14.

Steurs, M., Mulder, M. and Van Passen, M.M. (2004). A cybernetic approach to assess flight simulator fidelity. *AIAA Modeling and Simulation Technologies Conference and Exhibit.* AIAA 2004-5442, Providence, Rhode Island, 16-19 August.

Taylor, H.L., Lintern, G., Holin, C.L., Talleur, D.A., Emanuel, T.W. and Phillips, S.I. (1999). Transfer of training effectiveness of a personal computer aviation training device. *International Journal of Aviation Psychology, 9*, 319-335.

Taylor, H.L., Lintern, G., and Koonce, J.M. (1993). Quasi-transfer as predictor of transfer from simulator to airplane. *Journal of General Psychology, 120*, 257-276.

Tomlinson, D. (2004). Using speech recognition in ATC simulation to generate an interactive cockpit communications environment. *Flight Simulation: 1929-2029: A Centennial Perspective.* London: The Royal Aeronautical Society.

Tulving, E., and Thompson, D.M. (1973). Encoding specificity and retrieval processes in episodic memory. *Psychological Review, 80*, 352-373.

Ungs, T.J. (1989). Simulator induced syndrome: Evidence for aftereffects. *Aviation, Space, and Environmental Medicine, 60*, 252-255.

Welch, R.B. (2002). Adapting to virtual environments. In K.M. Stanney (Ed.), *Handbook of Virtual Environments*. Mahwah, New Jersey: Lawrence Erlbaum Associates.

Wildzunas, R.M., Barron, T.L., and Wiley, R.W. (1996). Visual display delay effects on pilot performance. *Aviation, Space, and Environmental Medicine, 67*, 214-221.

Witmer, B.G. and Singer, M.J. (1998). Measuring presence in virtual environments: A presence questionnaire. *Presence: Teleoperators and Virtual Environments, 7*, 225-240.

Zindhol, J.M., Askins, T.M., and Sission, N. (1996). *Image update rate can affect the perception of simulated motion.* USAF AMRL Tech. Rpt. No. AL-HR-TR-1995-0194.

Index